GLOBALIZATION AND THE WORLD OCEAN

GLOBALIZATION AND THE ENVIRONMENT SERIES

This AltaMira series publishes new books about the global spread of environmental problems. Key themes addressed are the effects of cultural and economic globalization on the environment; the global institutions that regulate and change human relations with the environment; and the global nature of environmental governance, movements, and activism. The series will include detailed case studies, innovative multi-sited research, and theoretical questioning of the concepts of globalization and the environment. At the center of the series is an exploration of the multiple linkages that connect people, problems, and solutions at scales beyond the local and regional. The editors welcome works that cross boundaries of disciplines, methods, and locales, and which span scholarly and practical approaches.

SERIES EDITORS:

Richard Wilk, Department of Anthropology, 130 Student Building, Indiana University, Bloomington, IN 47405 USA or wilkr@indiana.edu

Josiah Heyman, Department of Sociology & Anthropology, Old Main Building #109, University of Texas at El Paso, 500 West University Avenue, El Paso, TX 79968 USA or jmheyman@utep.edu

BOOKS IN THE SERIES:

1. *Power of the Machine: Global Inequalities of Economy, Technology, and Environment*, by Alf Hornborg (2001)
2. *Confronting Environments: Local Environmental Understanding in a Globalizing World*, edited by James Carrier (2004)
3. *Representing Communities: Histories and Politics of Community-Based Natural Resource Management*, edited by J. Peter Brosius, Anna Lowenhaupt Tsing, and Charles Zerner (2004)
4. *Globalization, Health, and the Environment: An Integrated Perspective*, edited by Greg Guest (2005)
5. *Globalization and the World Ocean*, by Peter Jacques (2006)
6. *Global Visions, Local Landscapes: A Political Ecology of Conservation, Conflict, and Control in Northern Madagascar*, by Lisa L. Gezon (2006)

GLOBALIZATION AND THE WORLD OCEAN

PETER JACQUES

PRESS

A Division of
ROWMAN & LITTLEFIELD PUBLISHERS, INC.
Lanham • New York • Toronto • Oxford

ALTAMIRA PRESS
A Division of Rowman & Littlefield Publishers, Inc.
A wholly owned subsidiary of The Rowman & Littlefield Publishing Group, Inc.
4501 Forbes Boulevard, Suite 200
Lanham, MD 20706
www.altamirapress.com

PO Box 317, Oxford, OX2 9RU, UK

Copyright © 2006 by AltaMira Press

All rights reserved. No part of this publication may be reproduced, stored in a retrieval system, or transmitted in any form or by any means, electronic, mechanical, photocopying, recording, or otherwise, without the prior permission of the publisher.

British Library Cataloguing in Publication Information Available

Library of Congress Cataloguing-in-Publication Data

Jacques, Peter.
Globalization and the world ocean / Peter Jacques.
    p.   cm. — (Globalization and the environment)
Includes bibliographical references and index.
ISBN-13: 978-0-7591-0584-3 (cloth : alk. paper)
ISBN-10: 0-7591-0584-7 (cloth : alk. paper)
ISBN-13: 978-0-7591-0585-0 (pbk. : alk. paper)
ISBN-10: 0-7591-0585-5 (pbk. : alk. paper)
   1. Ocean—Environmental aspects. 2. Marine ecology. 3. Marine resources conservation. 4. Marine resources development. 5. Globalization. I. Title. II. Series.
GC28.J33 2006
333.91'64—dc22   2005034553

Printed in the United States of America

∞ ™ The paper used in this publication meets the minimum requirements of American National Standard for Information Sciences—Permanence of Paper for Printed Library Materials, ANSI/NISO Z39.48-1992.

To Pamela, whose persistent labor at work and at home allowed me to finish school, which was something that she always believed in and never gave up on—for me. Thank you.

CONTENTS

List of Figures and Tables		ix
Acknowledgments		xi
1	Sea Mist and Salt Air for Sale	1
2	Global Environmental Theory, Oceanic Circles	17
3	Marine Political Ecology	39
4	Sustainability in the South Pacific	65
5	Sustainability in the Caribbean Basin	87
6	Sustainability in Southeast Asia	111
7	Connecting the Parts—Theoretical Connections	137
8	Conclusion	161
References		167
Index		185
About the Author		191

FIGURES AND TABLES

Figures

Figure 3.1.	Global Wild and Aquaculture Fish Catch, 1950–2000	43
Figure 3.2.	Global Wild Marine Fish Catch, 1990–2001	44
Figure 3.3.	The Degree of Global Fishery Exploitation	44
Figure 3.4.	North-South Wild Marine Fish Consumption over 30 Years	48
Figure 3.5.	Scale of North-South Marine Fish Production	48
Figure 3.6.	Scale of United States Carbon Emissions	59
Figure 4.1.	Change in Foreign Direct Investment (FDI) in the South Pacific, 1988–2000, with and without Australia	69
Figure 5.1.	Change in Foreign Direct Investment (FDI) in the Caribbean	93
Figure 6.1.	Change in Foreign Direct Investment (FDI) in Southeast Asia, with and without China	116
Figure 7.1.	Average Regional Foreign Direct Investment (FDI)	152
Figure 7.2.	Armed Conflict and Foreign Direct Investment (FDI) between Regions	153
Figure 7.3.	Armed Conflicts and Foreign Direct Investment (FDI) across Regions, 1980–2002	154

Tables

Table 3.1.	North-South Trade of Fish	47
Table 3.2.	Global North-South Wild Marine Fish Catch	47
Table 3.3.	Top 25 Total Carbon Emission Countries in 2000	55
Table 3.4.	Top 25 Per Capita Carbon Emissions in 2000	56
Table 3.5.	Bottom 25 Total Carbon Emission Countries in 2000	57
Table 3.6.	Bottom 25 Per Capita Carbon Emissions Countries in 2000	58

Tables 4.1.	Level of Globalization in the South Pacific	68
Table 4.2.	Corporations, Debt, and Assistance in the South Pacific	68
Table 4.3.	South Pacific Fisheries	70
Table 4.4.	South Pacific Coral Reefs	71
Table 4.5.	South Pacific Climate-Related Ocean Changes	72
Table 4.6.	Poverty and Government Expenditure in the South Pacific	74
Table 5.1.	Change in FDI in the Caribbean	92
Table 5.2.	Corporations and Debt in the Caribbean	94
Table 5.3.	Caribbean Fisheries	95
Table 5.4.	Caribbean Coral Reefs	96
Table 5.5.	Caribbean Climate-Related Changes	96
Table 5.6.	Poverty and Government Expenditure in the Caribbean	98
Table 6.1.	Change in FDI in Southeast Asia 1988–2000	114
Table 6.2.	Corporations, Debt, and Assistance in Southeast Asia	115
Table 6.3.	Southeast Asian Fisheries	116
Table 6.4.	Reefs at Risk in Southeast Asia	117
Table 6.5.	Southeast Asian Climate-Related Ocean Changes	118
Table 6.6.	Poverty and Government Expenditure in Southeast Asia	120
Table 6.7.	Armed Conflict in Southeast Asia, 1946–2000	122
Table 7.1.	Ecological Summary	138
Table 7.2.	Regional Political Summary	139

ACKNOWLEDGMENTS

FOREMOST, I AM GRATEFUL to Zachary Smith, my adviser and friend at Northern Arizona University who coauthored my first book with me, and without whom none of this work would exist. He picked me out of his class I was attending in 1997 for academia which was not my initial goal even though it was in many ways a lifelong ambition; he made this goal possible by working closely with me and providing advice that never set me astray. He has produced well over twenty PhD's with expertise in environmental scholarship. Imagine the impact of that.

I am also very grateful to Gabriela Kütting at Rutgers University for her early reading, discussions, and comments on this book, in addition to her participation in the panel mentioned below. She provided key advice in direction, and has been a friendly voice since the beginning.

I am also grateful to Elizabeth DeSombre, Yannis Kinnas, and Radoslav Dimitrov, who all provided feedback to chapter 2 at a meeting of the International Studies Association in 2004, coordinated by Gabriela.

Further, I am grateful for the comments of the CLIOTOP (Climate Impacts on Top Ocean Predators) Working Group 5, where the findings of this book were presented in December 2004. Vina Ram-Bidesi at the University of the South Pacific reviewed chapter 4, and I am especially grateful for her help. I am also very grateful to Kathleen Miller at the National Center for Atmospheric Research (NCAR) as well as Sam Pooley and Tanya Maiava at the University of Hawaii and NOAA Fisheries for their help with travel expenses. I also thank Katrina Rogers at the Arboretum of Flagstaff, Arizona, and Bruce Janz and Dwight Kiel at the University of Central Florida for their discussions on hermeneutics and knowledge. The political science department at the University of Central Florida, and Roger Handberg and Susan Devor in particular, supported the book by making the necessary travel easy and accessible. I am thankful for their support.

I would also like to specifically thank the three anonymous reviewers who provided attentive, insightful, and specific suggestions which have improved the text greatly. Finally, I would like to thank AltaMira editor Rosalie Robertson, whose affable personality made this process a pleasure.

All of the above people have helped me develop the ideas in this book; they are of course blameless for weaknesses and errors.

Sea Mist and Salt Air for Sale   I

THIS BOOK EVALUATES the sustainability of the World Ocean.[1] Several layers of confirmation indicate relatively recent and profound changes to the ocean structure, creating a need to understand large-scale trends to stem these changes. The transformations of the World Ocean are ones of which humankind has no collective memory, experience, or holistic knowledge, and there is significant evidence that the expanding global economy is closely related to these changes. This book is an experimental attempt to both conceptualize and understand these extraordinarily complex changes in a holistic, oceanic way. In this chapter, I introduce the initial reasoning behind this study, frame the problems, and describe the plan of the book.

Globalization, Ideology, and Ecological Imprints

Economic scale and movement have changed the human-ocean relationship by distancing affecters from effects and by alienating parts of the ocean from the whole. This means a broad understanding is needed to (begin to) comprehend the impact of economic globalization on the World Ocean. This section describes the reasoning behind these changes.

In the last several decades, trade all over the world has increased in volume and intensity—that is, trade has become more pronounced through a process now referred to as the economic aspect of *globalization* (Kim and Shin 2002). Globalization, like sustainability, is a contested term. I assume here that *globalization* has a range of definitions and that the term minimally refers to the "stretching of social, political and economic activities across frontiers such that events, decisions and activities in one region of the world can come to have significance for individuals and communities in distant regions of the globe" (Held et al. 1999, 15). This means that connections between people have an increased intensity, speed, and velocity in social, political, economic, and environmental arenas across regions (Held et al. 1999). As these connections become "universal," the maximal definition of the term *globalization* is satisfied (Kütting 2004); when I use the term I will typically be referring to the minimalist position unless I specify otherwise. However, I assume that if a system is global in the universal sense, as I theorize the World Ocean is (chapter 2), then changes will be only temporarily stationed at the regional level, since they will mix with the rest of the areas over time.

Chief among the politics of globalization is the role of the now worldwide network of capitalistic trade and the development of the "global economy." This economy has built upon various stages, starting with imperial expansion of European powers in the sixteenth century. In the last century, trade was common and economies were open until 1913, but this situation ended with World War I and remained closed and regional in scope until the end of World War II. The boom of the 1950s saw international, interregional trade open up again, only to increase since that time. In these hundred years, the world economy has expanded twentyfold. James Gustave Speth (2003) puts it this way: "It took all of history to grow the world economy to $6 trillion by 1950. It now grows by this amount every five to ten years. Since 1960 the size of the world economy has doubled and then doubled again" (2).

Within this last period of exponential growth of capital, there has been a distinct period of economic integration, which comes from an increase in international trade, migration, and capital flows—all of which have hit an "unprecedented" level since the 1980s (World Bank 2002, 23). I think of the period stretching from the 1980s to the present as the current phase of the third wave of globalization.

The 1980s was also the time that marked the ascendancy of neoliberalism, defined below, and the readiness of the United States and its allies in the countries of the Organization for Economic Cooperation and Development (OECD) to "assert their global power after a period of introspection in the 1970s" in order to open up the global South to transnational corporations (Brohman 1995, 134–35).

Similarly, in 1982 the Law of the Sea was finalized directly after U.S. President Ronald Reagan abandoned the treaty and unilaterally implemented parts of the treaty he liked (such as claiming 200 miles of the coastal zone). He simultaneously disavowed international wealth redistribution, which was found in the Common Heritage function in the treaty (described in chapter 3) and was incompatible with Reagan's neoliberal program. This neoliberal force, along with its underlying liberal tenets, has been so powerful and compelling that some have described it as the "end of history," implying that this kind of political organization is the final purpose of human existence (Fukuyama 1992). I forcefully reject this claim, but it is clear that along with the expansion and integration of markets and the global economy, neoliberalism has extended throughout the world. This phenomenon has been demonstrated by the 1990s, which saw the dramatic globalization of countries which formerly saw this expansion as imperialistic. We have now witnessed the integration of these former "second world" countries from the Soviet and Communist bloc into Western institutions like the North Atlantic Treaty Organization (NATO), the European Union, and the World Trade Organization.

Respected ecological philosopher Andrew Dobson provides a helpful discussion of the accompanying asymmetry to this expansion when he considers how globalization has changed citizenship. Dobson uses Castells for context:

> In a global approach, there has been, over the past three decades, increasing inequality and polarization in the distribution of wealth. . . . The poorest 20 percent of the world's people have seen their share of global income decline from 2.3 percent to 1.4 percent. . . . Meanwhile, the share of the richest 20 percent has risen from 70 percent to 85 percent." (Castells in Dobson 2003, 19)

Thus, globalization is not an even process of economic expansion and opportunity where everyone is connected and everyone becomes an equal part of a wondrous network of global invisible hands. Instead, while there are some opportunities for poor countries and their civic groups, globalization moves mostly in one direction. Global activist Vandana Shiva elaborates that "Through its global reach, the North exists in the South, but the South exists only within itself, since it has no global reach" (Shiva in Dobson 17). This does not mean that globalization is inherently "bad" and localization "good"; it means that historically, globalization has occurred to the privilege of some and at the expense of others. Nonsustainable trends are embedded in inequitable power relationships; thus, global material equity is necessary for curbing maldistribution and exploitation of resources.

Dobson rejects the more cosmopolitan belief that there is a reciprocal obligation of everyone to one another in favor of a distributional responsibility such as from North to South based on the materials produced and reproduced through asymmetrical globalization. This is a more sophisticated iteration of the material equity included in the Borgese Test described below. I take Dobson's (and Shiva's) point that globalization enables this connectivity through and within ecological spaces and budgets, and that sustainability requires benefits to be redistributed throughout transnational communities (Dobson 2003). It is worthwhile to reflect on the question "How much has changed for the majority of poor countries in the last fifty years, and in particular the last twenty years, in the face of Western 'help'?" and then to simultaneously ask, "What is the direction of ecology in this same last 50 years?" Minus a few exceptions, the promise and dream of "development"[2] for the global South has actually "produced its opposite: massive underdevelopment and impoverishment, untold exploitation and oppression" (Escobar 1994, 4) at the same time ecology has seen "structural" decline—that is, a decline of the frame and foundation of ocean ecology. Structure is important economically and socially as well and implies the same meaning of the larger frame and construction of a system where constituent agents and decisions are made, but which do not fundamentally alter the larger design. As a political scientist, I cannot count this situation as an accident, but instead a purposeful result that can come about only through disproportionate and asymmetrical structural arrangements of power—but from where?

While localities cannot escape some responsibility, poor localities have unquestionably been marginalized, and their ability to change their situation has been fraught with obstructions that originate from the colonial period. Much of this power has come from development discourses and projects which embody the ideals of what progress should be (through modernity), and this has then framed the reality that poor countries find themselves in when needing stabilization loans or making trading arrangements (Escobar 1995). This follows the various ghosts of modernity, now supported, re-created, and defended most by the ideology of "neoliberalism."

Liberalism is the central Western political theory, ideology, and political economy preferring a least restricted market, pluralistic competing political groups such as NGOs, various strong civil freedoms for individual citizens (e.g., of speech, religion, etc.), and a neutral State which affords procedural equity (procedures of the state treat everyone the same, e.g., in a courtroom) to all citizens and most agendas.

Neoliberalism is a reformed liberalism that places much more focus on the market aspect of liberalism and much less focus on civil liberties. Neoliberal policies focus on privatizing formerly public enterprises and industry; lowering social expenditures of the state (particularly those which tend to redistribute wealth); reducing or eliminating tariffs toward other countries; and creating a tax and physical infrastructure that favors industrial production and trickles down to lower classes to create economic growth and employment and reduce poverty (Friedman 1962). Neoliberal policies are not concerned with creating a social safety net, leaving this up to a robust economic growth, nor do they like regulatory environmental policies, which they prefer to leave up to the pricing of goods. This ideology is exported through trade and loan arrangements to other countries from the Western power elite, such as World Bank, the OECD, or individually through the United States, Britain, and some other European countries that have majority voting power in the World Bank and the International Monetary Fund (IMF).

Neoliberalism changed its focus from simple capital accumulation models to include the development of institutions. Evans (2004) sees these institutions imposed, such as through international finance institutions, which are Anglo-American-generated models. These institutional designs are all the same; he calls them "institutional monocropping," where the "best response to bad governance is less governance" (35). This arrangement is a fundamental problem with neoliberalism because it creates fewer limits on exploitation of people and natural resources, and places the profit motive of firms in a privileged position without any sense of citizenship mentioned above. In contrast, Evans proposes, along with scholars Dani Roderick and Amartya Sen, that the building of institutions should center around more direct and participatory deliberative democratic institutions where minority voices have more influence to stem exploitation.

Ironically, the Anglo-American set of institutions and countries never strictly employed neoliberalism themselves. It is well known that state involvement and guarantees (to differing degrees) of some civil rights and social welfare have been key elements in the building of stable industrialized affluent countries (Hettne, Inotai, and Sunkel 2001). The United States and the European Union (EU) consistently use state subsidies and protections, such as for agriculture—the primary area in which industrializing countries have a competitive advantage (Kütting 2004). For ocean politics, the Northern subsidies of fishing fleets are a source of overfishing and a prime example of a non-neoliberal policy, which is now being negotiated in the World Trade Organization (WTO). Nonetheless, Anglo-American countries demand lightning-fast change toward free markets and liberal democracy, without some level of democratic guarantees and social welfare. Evidence indicates that this can and has led to instability, violence, and ethnic genocide because these rapid changes create unequal market and political controls among factious rival groups (Chua 2003). This is not occurring only at the national level.

The Third World cannot compete against Northern subsidies. This problem was symbolized by a South Korean farmer, Lee Kyoung Hae, who committed suicide outside the 2003 WTO meeting in Cancun, Mexico, as a protest to WTO rules that allow agricultural protections from the free market (Vidal 2003). Protesters at this meeting numbered over ten thousand and hailed from more than thirty countries; they presented some recurrent demands, which "included protection from big business, abandonment of genetically modified crops in developing states, and no privatization of water, forests and land" (Vidal 2003).

Now, the world economy is growing at about 5 percent per year—the fastest in almost thirty years (International Monetary Fund 2004). This global economy is based on flows of energy, material, and capital. This flow is called *throughput*, and is used to sustain (and impoverish) groups within the population greater than six billion people. These energy flows start and end within natural systems. More throughput means more withdrawals and additions from and into natural systems. Therefore, the basis for connecting economic globalization to ecological decline is that current globalization expands the scale and intensity of throughput; this kind of growth is viewed as essential to progress and development within neoliberalism.

The inherent disconnect between resource decline and global economic expansion is hidden structurally by distancing, or "distanciation" (Kütting 2004). Capitalism in general, but in particular petroleum-based capitalism, creates expedited pathways for export and trade that become separated, or distanced, from their local meaning so that "Time-space separation disconnects social activity from its particular social context" (Kütting 2004, 33). This is related to what ecological economists have described in terms of ecological burdens remaining outside the pricing system as externalities which ecology and third parties eventually pay. Current globalization allows affluent populations to shift environmental costs through a global economy, and these populations are structurally permitted to live off of the carrying capacity of others (Kütting 2004; Muradian and Martinez-Alier 2001a, 2001b; Martinez-Alier 1995; Bunker 1985). When one fishery is depleted, the world economy can move on to the next fishery, structurally obscuring the problem because consumers are not dependent on local ecological budgets. Changes do not affect affluent consumers because these customers are not forced to care about the first depleted fishery, and in this way human-ocean relations have fundamentally changed with economic globalization.

Poor people of the world, however, still subsist and depend directly on local ecological functions. This does not mean that local governance is inherently "good," especially in terms of true global commons like the atmosphere and the World Ocean that require international governance. It means that local, regional, and global rules need to be accountable to their impacts. Thus, it may seem that distanciation implies a preference for local control of ecological impacts because as a practical matter distancing is harder to hide at the local level and there is an assumption that this transparency is important for creating accountability. However, any level of government that shifts accountability can create distanciation, even at the local level.

Defenders of neoliberalism contend that a healthy global free market will be able to account for these ecological problems through pricing adjustments that will change supply, demand, and substitutions, and that fears of open trade and its related issues are overblown, misunderstood, and unfounded in empirical evidence (see Burtless et al. [1998] for an excellent defense of economic globalization). Nonetheless, there is widespread international empirical evidence that this system creates severe problems in political empowerment, economic equity, and ecological decline—all key elements of sustainability. For example, in some cases, neoliberal policies to privatize commonly held resources, like coastal zones, actually *depend* on repressing civil-society members who continue to resist the encroachment of private interests, like shrimp-farming companies, into public goods like coastal mangrove forests. Obviously some privatization is beneficial, and there are indications that some privatization of fisheries can aid conservation (Jacques and Smith 2003); but the "monocropping" of privatization for the sake of simple economic growth has negative side effects that are not sustainable.

Further, inasmuch as this process incurs debt, global financial organizations facilitate the building of an infrastructure that has not markedly made improvements for the poor, but has imposed austerity programs and a reduction in the capacity of governments to actually govern economies and firms. This undermines the process of developing regional and national institutions aimed at sustainability; these institutions are needed to manage environmental problems that have come with industrial scales of extraction, production, consumption, and development—all of which have increased the scale and degree of throughput in transregional networks within the global economy.

Throughput that continually increases requires an unlimited system of energy and resources in order to continue indefinitely. However, ecological systems are limited by both the stocks and rates of energy and resources available. Economic use of resources in this way also increases ecological entropy. Georgescu-Roegan (1971), a founder of ecological economics, described the conflict between a growth-based economy and entropy. Industrialized societies have switched their agricultural economic base from solar energy, which has unlimited stock but limited rate of use, to hydrocarbon-based economic production that has finite stock and unlimited rate of use through coal and oil. This shift has produced machines that increased the ability to exploit resources. Consequently industrialization and growth have unfolded rapid ecological entropy where energy has been dissipated more quickly than had it been left within the natural cycles that produced it, for example, trees versus lumber. However, this less entropic energy prior to economic usage is not valued as much as the products and energy gained from the more entropic form in the short term. Thus, the incentive, without strong sociopolitical mores against such practice, is to overexploit resources toward unnaturally rapid entropic change, and this situation in turn underlies problems of carrying capacity, intergenerational equity, and material sustainability.

The current form of economic globalization mostly removes these sociopolitical restraints that may otherwise slow this move toward higher entropy in ecological systems.[3] Compared to capital accumulation within a neoliberal political economy, other important factors like ecological values, science, and human health become secondary and less potent to challenge this suicidal world system (Ridgeway 1996; Speth 2003; Mol 2001; Ostergren and Jacques 2002).

These warnings recur within the literature of sustainability, and in this book. Despite these red flags of danger, the extending reach of economic globalization continues to institutionalize the economistic and neoliberal perspective, which makes challenging the status quo more and more difficult. Although there are signs of hope and optimism, this book shows that the more economic neoliberal globalization becomes intensified in a region, the less sustainable the region becomes; neoliberal economic globalization is specifically undermining World Ocean systems.

None of the above should be taken as dismissive to economic activity or trade in general; rather, the above applies to the rules, logic, and rationality of neoliberal and economistic political economy. Sometimes these criticisms are framed as a "hatred" for capitalism or freedom (e.g., the polemics of Arnold and Gottlieb 1994). I criticize neoliberalism not because I hate capitalism or freedom;[4] I criticize economism, or treating all noneconomic values as secondary, for similar reasons as Paehlke (2004). Robert Paehlke has powerfully argued that placing economics above all else has undermined democratic efforts and subverts legitimate social and ecological programs around the world. Despite this, for some time now, "Economic considerations overwhelm all else. What might be called 'economism' is triumphant" (viii).

Clearly, trade is important for sustainability. Historically sensitive research on island sustainability and survivability shows that contact with other groups is especially important for the islands' isolated economies (Matthews and Gaulin 2001). However, I assume that economics should be controlled and governed by society to avoid the above problems. Further, I assume there are important differences found in scales of production and consumption, feedbacks to ecosystems and human groups, and the logic of sustainable resource use and ecological impact between political economic choices, such as neoliberalism, democratic socialism, communism, sustainable development, or traditional economies.

Also, criticism against neoliberal globalization does not mean that state-centered regulation is the automatic rejoinder. Many green theorists, such as ecoanarchists like Murray Bookchin (1982), reasonably see the state as a coconspirator with capitalism against sustainability. In addition, much of green thought—indeed, conservation biology—is seriously looking at the convergence of sustainable solutions at the local level in concert with industry, communities, NGOs, scientists, and other stakeholders (e.g., Norris et al. 2002; Stolton and Dudley 1999). These projects demonstrate that there are political solutions to governing apart from state-based regulation, though regulation should not be thrown out of the menu of choices.

For example, a tuna boat captain devised *backdown*, an innovation in reducing dolphin mortality in tuna bycatch in the Atlantic. Backdown, which involves backing up a seiner boat in a U shape to allow dolphins to escape, has been partially responsible for reducing tuna-related dolphin mortality by 99 percent (Norris et al. 2002). Of course, prior to the activism and pressure from environmentalists in the late 1980s, captains may not have been asked to consider the matter of dolphin mortality. Now, through the Agreement on the International Dolphin Conservation Program (AIDCP), environmentalists, scientists, and fishers through Inter-American Tropical Tuna Commission (IATTC) have created an institutional setting where the dolphin bycatch is limited and improving. It is worth noting, however, that it seems that the pressure to save dolphins has created other changes in fishing behavior, such as using floating attraction devices (where fish come to the device and are caught). This fishing method has increased the bycatch of many other species, perhaps by a whole order of magnitude—indicating that there needs to be further adaptation to avoid the entrapment of other species like sharks, billfish, and undersized tuna (Norris et al. 2002).

Questions and Problems of Ocean Sustainability

The above discussion does not reveal anything new, given that this debate has been going on for some time. My goal is to turn the discussion specifically toward the World Ocean. Among the questions addressed in this book is the connection between the above economic and ecological changes as they relate to the World Ocean.

In order to treat the topic, I take an openly interdisciplinary approach so as to connect scientific findings and observations with social and philosophical observations to construct a holistic picture of the World Ocean and the trends related to sustainability. I do not create new data, but interpret and shape existing data with observations that had previously been distinct. This is not an elegant and simple collection of work drawn neatly together; for better or worse, the issues at stake are numerous, complex, and messy. Some of this complexity can be strategically avoided if the reader is interested only in the empirical descriptions of the regions, found in chapters 4, 5, and 6, which come with only specific interpretations at the end of each respective area.

However, the intent of this book is more than just to describe ocean conditions. The intellectual contribution of this book is to connect several disparate but vital areas of knowledge. This book connects marine and atmospheric science to each other on specific topics, specific global environmental changes to likely anthropocentric causes and politics, and the development of oceanic institutions (defined very broadly here as patterns of governance/decision making through rules and roles) to cope with and redirect these effects. In order to make sense of these connections, we must engage theory. As such, I engage a set of epistemological choices for framing globalization and the environment to use in comparing against Elisabeth Mann Borgese's ideas of sustainability.

The Borgese Test

Most simply, sustainability is the convergence of improving social, political, economic, and ecological conditions (Goodland 1995). In what I am calling the "Borgese Test," I specify what this means for the ocean.

Borgese was a political scientist and international-law scholar at Dalhousie University in Canada, as well as a strident advocate for the ocean and human justice. Moreover, along with her colleague and Maltese delegate, Arvid Pardoe, she was a sincere advocate for the "common heritage of humankind" (chapter 3) provision within the Law of the Sea, which intends to distribute resources from the high sea soil to the poor and the cause of international development.

Borgese wrote several important documents, but *The Oceanic Circle* (published in 1998) was among her most important contributions. *The Oceanic Circle* describes sustainable ocean governance, and she uses Gandhian thought to make her case for saving the seas and people who are dependent on them. Nonhierarchical and nonviolent social relations should inform local management of resources with global cosmopolitan consciousness (knowing that what one locality does affects and has a responsibility to others). This is what she meant by making "oceanic circles," which she believed reflected the actual organization of the ocean itself.

Her plea is for radical democracy, nonviolence, and material equity, which are essential to nonhierarchical relations. Importantly, global equity means that no one is deprived of basic needs. It does not imply equal shares of goods or wealth. Further, she argues that this social change can occur as societies develop a deepening relationship with the global ocean. This requires grassroots empowerment to make global governance accountable; nonviolence; knowledge of interdisciplinarity; and global North–South equity, some of which is articulated by Gandhi in his poem "Oceanic Circles" (Borgese 1998). Resources should be comanaged through decentralized democratic authority, with the aim of using and improving ecological productivity and function, coordinated with national, regional, and global governance (part of "comanagement"). North-South equity implies that material conditions of the industrialized countries should not impoverish poor countries. Interdisciplinary science is used to avoid hierarchical knowledge-based power to approach complex environmental *problématiques* with "solutiques," or holistic global solutions. I impose on this definition the expectation that sustainability is a set of long-term processes, instead of an ideal which can easily become a form of authoritarian design from above; I believe Borgese would find this acceptable (see Lee 1993; Capra 2002). In sum, sustainability is the evolution of nonviolent governance accountable to multiple levels of human organization ensuring global human material equity and productive ecologies through interdisciplinary knowledge.

I will refer to this definition of sustainability as the "Borgese Test." One region cannot live unsustainably without endangering the livelihoods of the rest. This is captured in Borgese's ideas of North–South equity. If the North lives off of and undermines Southern ecology while the South lives in squalor, social and ecological sustainability is endangered, to varying degrees, around the world. Also note that sustainability could be defined as simple stocks and flows of energy and material, but I use Borgese's ideal because it includes the politics of justice that determine human use of stocks and flows. Steep social hierarchy, often empowered through violence, allows for ecological resources to become concentrated and overexploited, reinforcing the hierarchy and flow of resources and potentially triggering scarcity and more violence, ad infinitum until the system reaches impenetrable limits forcing rearrangement. Thus, distributive and nonviolent justice is fundamental to a sustainable world.

We are not building Borgese's hopeful oceanic circles, and global ocean sustainability is, if anything, slipping farther and farther away. Neoliberal globalization has increased hierarchies at the coastal level, and I show that along with increased economic globalization comes increased armed conflict. Violence and neoliberal economics seem to be globalized together (Chua 2003).

In places like the South Pacific where the oceanic circle seemed to have the strongest cultural salience, the circle is being challenged by economic globalization, specifically through the hubs of Australia and New Zealand. From the Borgese Test, we can see, through several theories employed to interpret this test, that instead of creating a more sustainable world, we are mostly creating a more balkanized and abusive one that is causing globalizing environmental changes, despite progress in the areas of interdisciplinary locally based science and more regionally connected grassroots activism.

The Importance of the World Ocean in Global Studies

The World Ocean, as a singular connected body of regional oceans, has been essential for globalization from its earliest inceptions because it allowed for powerful societies to cross regions with relative speed. The earth, after all, is over 70 percent water. Almost all of this water is connected in one continuous body, the World Ocean, which itself is connected to the land and atmosphere that complete the whole Earth system.

Less than 1 percent of the earth's habitable space is on terra firma, making almost all of the space in which life inhabits the planet water within the ocean (Borgese 1998; Costanza 2000). And perhaps most important is that *all* earthly life began in the World Ocean, and *all* life continues to physically depend on a few key functions of the ocean, even the lives of those people who live in landlocked areas. In this sense, the World Ocean is a critical ecological system that underlies the security and well-being of all organisms that depend on oxygen and a stable temperature (this is discussed in more detail in chapter 3).

Our connections through the water are universal. If these connections are fostered in sustainable and pacific terms, our existence will be enriched, protected, and enlightened. Broken or thrown out of balance, slight conditions may change in the oceans and change the way people live and die. Perhaps temperatures rise a bit in the Pacific, and within a few months, thousands of people die as a result of massive mudslides in Bangladesh at the same time that forest fires rage in the western United States and fisheries all of a sudden disappear temporarily off the coast of Chile. Unfortunately, humanity's ability to make these connections has not matched our ability to alter them. At the same time, global political structures determine human organization often in contrast to well-established knowledge. Thus, knowledge and governance need to evolve together. Some of these global structures even work to disaggregate our knowledge and ability to recognize these problems (see the section on mechanization in chapter 2). I attempt to draw these elements together in this book-length thought experiment.

Only now is marine science building specifically global data collection, such as through the Global Coral Reef Monitoring Network, the Global Ocean Observing System, and the World Ocean Circulation Experiment. At this time, social science is also more cognizant of globalization in economic, anthropological, political, and other human terms. Putting these global observations together now may provide key information about the future of not only ocean health but also human survivability. This does not mean humanity is doomed if the ocean changes—it is changing all the time, particularly from a geologic time frame. Rather, it means that human lifestyles and choices are altered alongside oceanic decline and change. I do, however, believe this decline is existential in nature for humans, because these changes may alter the meaning, if not the orientation and fecundity, of human life as we know it.

The first frontier of this book is to reconstruct theories of the global in order to understand the meaning of regional changes and conditions. I believe the ocean is by definition a global element. By a conservative count there are at least three universally global ecological systems: the World Ocean, the global gene pool, and the atmosphere. These are global ecosystems because they span across all geographic regions, and even create the conditions for other ecological functions.

All three of these global subsystems are within the larger Earth system and directly and indirectly affect one another. Analysis of these changes will be at best imperfect, but the practice is necessary if we are to even estimate the process of ecological global changes and the effects on human beings. I state at the outset that I assume the claims within this book are contextual to our time period, subject to deep wells of uncertainty, and apt to change. Trends away from sustainability do not mean conditions in regions or the world are determined. Political and creative solutions abound, a global environmental consciousness is a reality (Dunlap, Gallup, and Gallup 1993), and life itself is creative.

That being said, time is advancing, and choices are narrowed as systems are simplified; and as we use nonrenewable resources like oil without using them to create substitutions, we reduce our capacity to act in the face of problems that could otherwise be solved. This warning is consistently articulated in the sustainability literature; the question is, will human relationships be able to reorganize according to evolving principles of sustainability, or will the siren Cassandra call the ship's captains to a provocative shore?

Organization of the Book

In chapter 2, I discuss global theory and how I use theory to understand global change and sustainability. In chapter 3, I discuss some basic issues of ocean ecology, politics, and policy that are necessary to understand before moving forward. In this chapter I also provide some context for the major ecological systems I have chosen to focus on, and discuss how these systems are thought to connect to one another. Chapters 4, 5, and 6 discuss conditions found in the respective regions. Chapter 7 brings together the findings of the regional conditions into both a global theory and an understanding and interpretation of World Ocean sustainability, using the Borgese Test as the benchmark. Chapter 8 explores conclusions from the research about ocean and human security. Each section will now be discussed in a bit more detail.

Sustainability is a contested notion complicated by subjective political intentions and bias. As a way to derive more compelling findings, this book uses three theoretical choices for making sense of the empirical descriptions—complex systems theory, hermeneutics, and critical theory. These theories provide a variety of perspectives about global oceanic phenomena. Using different viewpoints allows the opportunity for corroboration and consilience between shared interpretations. Discussion of these theories is covered in chapter 2, with conclusions in chapter 7.

These changes in the World Ocean have profound importance. Global ecological changes have implications for the well-being of humanity as a whole, though some people are more affected than others. Those who are more touched are mainly the politically disaffected such as the poor and women (who actually make up about 70 percent of the world's poor) (Tickner 2004). Deciding how sustainability occurs or does not occur is a political struggle, and choosing the model of sustainability is part of this struggle. Some, as in the Brundtland Report (World Commission on Environment and Development 1987), look to concepts of "non-declining welfare" for present and future generations. This is not good enough. The majority of people on the planet need improving welfare. Nondeclining welfare for those who experience poverty-as-deprivation (not just a lack of money), preventable disease, political persecution, and regular social abuses as a matter of the status quo is not sustainable any more than is declining welfare for the global upper and middle class.

As noted above, Borgese places a great deal of emphasis on the grassroots emergence of institutions to deal with the World Ocean; these institutions will take their lead from metaphorical messages of the ocean itself. The most basic demand of these institutions is that the institutions establish nonviolent power relationships with other people and nature, and that peaceful uses of nature provide equitable economic conditions and basic life provisions for everyone, while protecting organisms, habitat, and ecological processes that underlie the economic process. These are the basic boundaries for sustainable institutions.

From looking at the institutions dealing with regional ocean changes, I have found that most do not fulfill this "Borgese Test" and that most formal or informal institutions that do fulfill this demand are traditional local practices; consequently, regionally and globally there are serious gaps. These gaps provide a direction for practical policy and political analysis, as well as clues about political causes of unsustainable conditions and trends.

Political connections to each of the global ecological changes are important to trace. If there are coherent trends in any of the global ecological changes (and there are some), then we can start from the assumption that these changes are not inspired by purely local phenomena, but occur as local phenomena within a global structure and design. For example, I trace the influence of global financial institutions on fisheries and coral reefs, to the extent that this information is available. I also describe the regional responses to these changes and their ability to alter decisions from global power structures. In some cases, as in the Caribbean, there is a concern for the ability to correct global power structures from a relatively disempowered region.

Some local resistance groups and NGOs are making regional and occasional global connections, and there is evidence that a growing global environmental civil society is emerging, as Paul Wapner (1996) suggests. But this is not without potential contradictions since as NGOs use global pathways created by a powerful minority, there is the potential for these struggles to become a Trojan horse.

Most globalizing environmental groups operating in poor countries are not from poor countries, and many NGOs are caught fighting local impacts instead of regional or global causes. The exceptions are rare Southern organizations like Third World Network (Malaysia). Most global NGOs are Northern NGOs like WWF. Thus, we are forced to ask tough questions about the impact of these organizations. Are these Northern NGOs actually facilitating the economic globalization that is behind so much of these declines? No doubt many, such as Oxfam, are admirably fighting against these declines. But the fact remains that the globalization of the World Ocean is being driven from primarily one side (centers of material power in the North) with some exceptions like CP Group of Thailand or the fishing fleets of China. Global and regional environmental NGOs have the potential to put a "green" face on globalization, such as through partnerships with firms to create marine protected areas (MPAs). But, land donations should not be traded for cozy relations with firms at the expense of accountability for firms and States. Co-opting local politics, then, is a danger Northern NGOs must avoid if they are to remain legitimate. In some cases this will mean making very hard choices about having more money that may come from firms or governments and that could be used for justice or conservation.

Specific trends of global socioecological changes come from three important tropical regional cases detailed in chapters 4, 5, and 6: the South Pacific, the Caribbean basin, and Southeast Asia. I chose the specific regions based on their preponderance of island states and their apparent regional identity based on formal regional cooperative agreements. Further, I chose these regions because they all had coral reefs, and all together represent about 70 percent of the world's reef area. Reefs are important indicators of sustainability of ocean and world ecological systems because they are among the richest areas of life in the world; they are undoubtedly one of the biosphere's most important areas and are crucial to global and ocean socioecological sustainability. Thus, all of these areas consider themselves to be regions with varying levels of international commitments, and these regions are forced to deal with and acknowledge the ocean in these commitments.

Within each regional chapter, I focus on a particular set of politics as it relates to ecological changes. First, I trace the role of global finance and trade in fishing, coastal development projects, and the production of knowledge. Second, I compare regional international institutions. I specifically describe the rhetoric of sustainability in each formal regional organization. In addition, I include sketches of informal institutions; this approach is consistent with my broad definition of what an institution is. I define institutions as broad patterns of governance found in rules and roles, including nontraditional concerns for institutions such as gender equality and poverty, because they reflect interpersonal politics that inform more formal structures, and probably resource use in the area. These aspects of the regional institutions are then compared in the concluding section within the regional chapters.

In sum, my purpose is to understand, through three theoretical lenses, ocean sustainability in three regions in a globalizing world measured against the criteria given by Elisabeth Borgese. These regions all have coral reef loss, fishery changes, and sea level rise as a part of climate change. I find that many of these ecological changes can be understood in terms of global political economic conditions. None of these regions reasonably pass the Borgese Test for sustainability, which means that current economic globalization is by and large not sustainable for the World Ocean.

Notes

1. See chapter 3 for a discussion on why I refer to the marine environment as the "World Ocean."

2. I put *development* in quotation marks here because it is inconceivable that designing modern societies on unsustainable organizational patterns is "development," not to mention progress.

3. There is a movement within the World Bank and other institutions to adopt green policies, but these policies are still based on infinite growth and neoliberal economic principles (Kütting 2004).

4. Such criticisms are rhetorical nonsense and should be identified immediately as crude ideology.

Global Environmental Theory, Oceanic Circles   2

WHAT IS "GLOBAL" and how do we know when something is really global as opposed to something which is confined to a particular locality?[1] What is driving global changes? What is the meaning of these changes? These questions are increasingly important in light of increasing ecological degradation.

The purpose of this chapter is to describe the role of theory in this book and explore the theories employed here. In order to do this, I first briefly describe the important aspects of three epistemologies that relate to epistemological holism. Then I describe what I believe to be the most appropriate level of analysis for global study, the region. I then look to all of the three epistemologies for not only multiple frameworks, but for frameworks that can be integrated with one another into a viewpoint that includes complexity and information connectivity, subjective interpretation as a basis for understanding, and sociopolitical critique.

Within ecological science, there is the expectation that parts of ecosystems interlock and depend on one another in a complex web of life. Given this interdependence, it is essential to explore and experiment with frameworks of analysis for the global whole to understand, even if marginally, what impact our localized changes have in relation to the rest of the biosphere. As such, there is increasing alarm about the changes occurring in the World Ocean; however, these changes are not easily understood in the context of deep and complex relationships that extend from one of the most basic life forms on the planet (plankton) to the entirety of Earth's atmosphere. While there has been a great deal of concern that the global condition of the world's oceans is deteriorating, there are few epistemological applications of global theory to provide a framework for systematic analysis. Such systematic analysis is fraught with subjectivity and bias, and rather than attempt to hide this bias, as does Enlightenment-type science which claims pure objectivity, I have chosen theories that deliberately include the subject as part of the knowledge building that occurs. My primary biases are toward material distributional equity, nonviolence, and ecological protection; but these biases do not preclude other imaginations nor are they unjustified, as described in the Borgese Test above.

Taylor and Buttel (1992) argue in their article "How Do We Know We Have Global Environmental Problems?" that "we know" we have global environmental problems partially because "scientists and political actors jointly construct them in global terms," and "in part, because we act as if we are a unitary and not a differentiated 'we'" (406). I do not believe it is actually as simple as this. I believe there is a global environmental problem because it is constructed in our imagination and has a corresponding potential in physical, intrinsic reality.

First, it should be noted that there are powerful counterconstructions against global environmental problems (e.g., Lomborg 2001 and other environmental skeptics). And, while a socially constructed understanding of the natural world implied by Taylor and Buttel is an important and potent component I do not dismiss, there is an equally important physical reality that is not socially imagined. Oceans predate our imagination and certainly our ability to socially construct anything, and they continue to operate within, around, and above human notions of reality. As much as life originated in this ancient ocean, it constructed us more than we construct it.

What we know about marine conditions is both constructed and apart from construction. John Searle (1995) points out that there are some things that are observer related and some that are intrinsic. The ways in which we engage the ocean as humans, such as through its economic and cultural functions, are observer dependent. Constructing the ocean as a whole as I do is an observation dependent on the subject, much like the 1960s holism found in the "age of ecology," where nature itself was framed as interconnected, spurred on by the observations of Rachel Carson (Nash 1990). This is what Taylor and Buttel argue for; I agree, but this social construction is only a portion of the reality. Like ecology itself, the ocean needs to be available for global framing, and this availability is not observer dependent.

That the ocean water is globally becoming warmer while it is rising is intrinsic to the ocean itself. That the biophysical ecology of the world is available to be described through material, biological life, and energy that flow interdependently between ecological regions and spaces is intrinsic to what nature is, even if our representations and understandings of such are dependent on subjective imagination.

To the extent that the observer-related and intrinsic realities meet, albeit tenuously and often through abstraction, we are able to represent something that is true, even if it is fleeting and contested by other representations. It is, in other words, possible, even if it is not essential or natural, for human-dependent subjectivity to peer through distorted lenses at real global environmental problems. Over time, I expect this representation will have dimensions of corroboration, refutation, and modification so that human understanding of global environments is possible through the pragmatic limitation imposed by consistency (Hayles 1995). Ecology as intrinsically interrelated energy-biology-matter is one such understanding that has been presented through subjective consistency (Soulé and Lease 1995).

Therefore, I believe it is reasonable to say that even though all knowledge is filtered through complex social prisms, this subjectivity does not mean knowledge itself, or my proposed global scope of inquiry is necessarily unjustified or capricious. There is a global ecological setting, just as there is a unified human "we," if only that "we" all share a space in the human family with similar ecological and biological needs, and the division of people into subgroups is worthy of critical scrutiny if not outright rejection (Hawkins 2002). Also, the local and specific (as opposed to the global and holistic) scope of inquiry is by no means ignored since most work in academia is highly specific—so much so that there may be an overemphasis upon the microlevel, permitting a fallacy of misplaced concreteness if it is unable to construct a holistic understanding (Daly and Cobb 1989).

Why is it appropriate to construct a global environmental epistemology specifically around the World Ocean? To answer this question, I defer again to Elisabeth Mann Borgese, who wrote in 1998 (the United Nations Year of the Ocean) about "oceanic circles." She said we need to take our epistemological cues from the ocean itself. Doing this allows the ocean to become a platform not for violence and exploitation of others, but a platform of nonhierarchical relations. She hopes that the ocean will bring people together through cooperation and harmony. In other words, she wants us to "think like an ocean" in the same way that Aldo Leopold (1966 [1949]) instructs us to "think like a mountain":

> The medium itself, where everything flows and everything is interconnected, forces us to "unfocus," to shed our old concepts and paradigms, to "refocus" on a new paradigm. Fundamental concepts evolved over the millennia on land, like sovereignty, geographic boundaries, or ownership, simply will not work in the ocean medium where new political, legal, and economic concepts are emerging. (6)

Borgese guides us to the preamble to the Law of the Sea, which states, "[T]he problems of the ocean space are closely interrelated and *need to be considered as a whole*" (Borgese 1998). I hypothesize that along with this globalization of human activities, there is a concurrent globalization of human impact on nonhuman nature, which Held and others describe as the "catastrophe in the making" (Held et al. 376–412). Specifically, I suspect this globalization of human impact to be particularly evident in the World Ocean.

As noted in the introduction, the World Ocean is a global ecological system connected in one continuous body. It is worth restating that less than 1 percent of Earth's habitable space is on land (Borgese 1998; Costanza 2000). This major ecosystem has, since the days of the first empire, been the highway of powerful nations. The World Ocean has been used for transcontinental transport and trade, such as the infamous sugar-cotton-slave triangle in the first wave of globalization. At this time, Europeans took slaves from Africa to North America, where the Europeans then forced them to grow sugar and cotton, which was then shipped to Europe for manufacturing. Today, no less than 60 percent of global trade in value is conducted through container ships, and this is expected to rise to 70 percent by 2010 (Coulter 2002). Furthermore, commodities from the ocean make up part of this globalizing. Fish are caught in the South Pacific, packed in Asia, and sent to North America using flags from yet another country, and workers from yet another. In fact, Theodore Bestor (2000) describes how Atlantic bluefin tuna landed in Maine are purchased at the dock by Japanese buyers, who then process and ship the tuna to fish markets in Japan. Among the markets that receive the tuna is the world's largest, the Tsukiji, which sells the fish to sushi chefs in such places as New York or Hong Kong. Thus, the ocean has been and continues to be one of the primary stages for economic globalization and global ecological changes, and in this way is an important issue to consider within this book series, *Globalization and the Environment*.

Interpretation, the Whole, and the Role of Theory

The determination of distinct cause and effect, as it is specifically inferred from semiexperimental social science theory, is not the project in this book. In other words, this work does not employ theory to measure the explicit and distinct impact of a set of independent variables on changing a dependent variable. Instead, the role of theory in this book is to provide an interpretation and understanding of a larger set of phenomena which are not separated from one another for their independent effects. This kind of approach would, in fact, be rather incompatible with at least two of the chosen theoretical frameworks, complex systems theory and hermeneutics, which do not allow for prediction or the notion of *ceteris paribus*, described below. Therefore, the role of theory here is to frame an understanding, a reading, and a critique of ocean sustainability.

Interpretation of global ocean systems is one aspiration of this project, but the direction and content of interpretation depends on underlying values and assumptions. Rather than stick to a static set of values and assumptions, this project takes an openly experimental approach to global ocean system interpretation in order to find a more compelling frame or combination of frames that will help make sense of life-changing ocean conditions.

Specifically, three epistemological perspectives are explored here: complex systems theory, hermeneutics, and critical theory. Each of these theories has been chosen for its potential to bring meaning to globalization and the World Ocean transformations. Complex systems theory has the potential to provide global understanding through its ability to see connectivity and informational relationships within a network that is always changing and hopelessly complex. Hermeneutics, an ancient form of interpreting sacred texts, has the potential to sew together narratives into a global understanding that is comprehensive. Critical theory was chosen for its social deconstruction and critique of political economic systems—which it usually frames in global terms formed by the first wave of colonial expansion and continuing through neoliberal forms of economic penetration. Thus, these three views of the world provide the basis for interpretation and understanding on a global level. I will now discuss some barriers to global theory found in the Enlightenment legacy of mechanistic thinking.

Parts and Pieces: Mechanistic Epistemology

Mechanistic thinking is premised on the idea that individual replaceable cogs can adequately explain the whole. If this is in reference to nature, then the cogs are organisms that lose their inherent value and there is little violation in exploiting them because the relationship is an instrumental one defined by the needs of the user. Mechanistic epistemology is an obstacle to understanding global phenomena and meaning because it is blind to constitutive relationships.

Constitutive relationships are relationships that make up a system where the joined pieces are more meaningful than their isolated individuality. These relationships create all theories of the global, but the nature of how these relationships change is based on the theoretical outlook. Thus, constitutive relationships make a group of parts more than their simple sum, and looking only at the isolated parts will, by definition, miss the dynamics of the system created by these connections. This implies a complexity so deep that humans' interaction with, in particular, large ecosystems like the World Ocean may be sustainable only through pragmatic experimentation (Lee 1993).

Perhaps a good example of a system with constitutive relations is a family.

People in families are indeed individuals with their own lives, and it is important to social scientists to understand individuals to the best of the scientist's capacity. However, the family is responsible, to at least some degree, for who individuals become and how they act. The family is not *entirely* responsible for who the individuals become, and the individuals are rarely, if ever, entirely independent of their family. Nonetheless, people in the family become a part of one another, making the family something sometimes intangible, but constitutive. Concern over only individual units, such as individual people, is important, but it necessarily blurs the importance of the systems and subsystems of which the individual is a part—language systems, ethnic and historical subsystems, religious creeds, geographic subsystems like watersheds, and in our case, ocean systems. This concern over the individual requires a certain rationality that focuses on the interests and behavior of individual parts. Similarly there is rationality within global theory that focuses on the interests of the whole and the system. And, like the family, the human and ecological systems examined here are alive and have their own inherent value so that mechanistic exploitation of the ocean will lead to dysfunction just as exploitation of a family or family member leads to the same.

Work that ideologically favors these systemic interests is sometimes labeled "globalism." However, this designation is really not specific enough to be helpful. Nearly everyone would support the globalization of peace and prosperity, though activist Arundhati Roy is well known for saying that the "only thing worth globalizing is dissent." However, along with this she says we need a sharing of spirit and community—across the world. This implies that other things are worthy of globalization:

> What we need to search for and find, what we need to hone and perfect into a magnificent, shining thing, is a new kind of politics. Not the politics of governance, but the politics of resistance. The politics of opposition. The politics of forcing accountability. The politics of slowing things down. The politics of joining hands across the world and preventing certain destruction. In the present circumstances, I'd say that the only thing worth globalizing is dissent. (Roy 2002, online)

This sentiment of resistance and common spirit implies that the tone and structure of this new politics are important because the peace and prosperity cannot arrive through authoritarian rule that forces these conditions. Nor can the peace and prosperity come with a short time horizon that ends because of ecological changes forced by exploitation. I am a globalist in the sense that I do see humanity in a common lot; however, this lot has been purposefully segregated and cosmopolitan ("we are all in it together") sentiments do not ring true while the majority of people in the world suffer important deprivations while a minority dine elegantly. Consequently, I assume that because we are in a common lot, this extreme difference in well-being is undesirable, and that moderating the extremes of economic globalization for the middle ground of economic well-being is morally right (see Conca 2001 for a discussion on "sustaining the middle").

When we take a global frame of reference, we are forced to place ourselves and our subjectivity within the system, because there is nowhere for human subjectivity to hide, and human interests become tied into the interests of the whole. This is because if I am thinking of the whole, I am included within the configuration, and it is implausible to think that there is a viable mechanical distance between the subject and objects of inquiry. Overtly dissolving mechanistic claims of distance and objectivity are important, since many philosophers have indicated this line of attack is often used to conceal exploitation of nature and society. A mechanistic epistemology breaks the pieces into distinct units and studies them apart from the whole, like cogs in a machine. The object of inquiry becomes the parts, not their relationships, which may be equally or more important.

What are the pieces that are obscuring the whole in ocean politics? Perhaps the most dizzying factor is that of the single-species fishery catch statistics. So much of our current understanding of the state of fisheries is made up of these numbers. This is problematic for several reasons. First, the number of fish caught does not and cannot indicate the health of a fish population because the precise number of the populations are not known, and are very often estimated based on those very catch levels, as a proportion of effort used in catching the fish. This is referred to as "catch per unit of effort." An increase in effort but not in fish catch implies a lower population.

Second, this means we are relatively limited in the knowledge about fish populations outside what people try to catch. Because about 75 percent of the world's marine fish focuses on 200 species, or about 1 percent of existing known species, the knowledge that is not captured in single marine fish catch statistics is startling (Holmlund and Hammer 1999). The fact that fish catch has risen tremendously since the 1950s (by about 300 percent) gives the impression that fish stocks are fine. This does not say anything about the structure of these fish, such as how much of the catch is top predator and how that proportion has changed over time, nor does it say anything about the state of marine biodiversity in general through these increased catches. Despite the fact that there have not been studies that justify single-species fish statistics as a measure of how much catch can be sustained, this is the primary method by which fishing policies are made (Earle 1995; Jacques and Smith 2003).

Even more menacing is the fact that fishing policies based on these single-species catch trends fly in the face of increasing ecological change. Loss of mangroves and other important coastal destruction, as noted in chapter 6, is disconnected from future policy on fishing levels, despite the importance of mangroves for fish nurseries. Likewise increased pollution levels from urbanizing areas in both the Caribbean and Southeast Asia. Likewise for climate changes for any of these regions. Thus focusing on the single-species catch per unit of effort is terribly inadequate to set sustainable fishing levels.

As a way to conceive of the ocean system without making the above kind of critical error (some of which cannot be avoided given available information), I have conceived of the World Ocean system as an amalgam of material, energy, and life that is functionally integrated. The material of the system is the water column, which contains heat and kinetic energy and the coastal zone that phases the terrestrial into the marine. Life in the system is seen through coral reefs and fisheries. I think of the coral and the fisheries as communities that function within these material settings, but changes in either can impact both. For example, changes in the coastal zone impact fish populations, which then change the diversity and kinds of plants and animals on the reefs. This specific functional relationship is seen in all of the regions because overfishing is contributing to declining reefs. The energy in the system—the heat and currents and waves (not to mention the chemical energy not discussed here)—impacts all of these sections.

Thus, mechanistically breaking the "object" of knowledge into pieces fundamentally distorts our ability to see "reality" and empirically and morally understand the world around us. Ecofeminist Carolyn Merchant has famously argued that changing from an organic model of the planet—one that promoted the image of the earth as a single living organism—to a mechanistic one was the beginning of industrialized environmental degradation because there was no ethical obligation to dead, discrete cogs in the earth (Merchant 1980).

However, one does not need to be an ecofeminist to argue in favor of holistic epistemologies. Raymond Holder Wheeler argued as early as 1936 that science is cyclic in its focus on the whole and on the part, what he terms vitalism/organicism versus atomism/mechanism. To see this cycle, Wheeler must use a holistic analysis:

> In order to see clearly, why science is now turning *organismic*, it is necessary to look at history as a whole. Such a perspective shows us that the main problem of science, any science, always has been to solve the part-whole

relation, the problem of the many in the one, of pluralism and unity, of permanence and change, *of the role played by the part in the whole.* (30; emphasis in original)

The possibility of holism is precluded by the mechanistic approach to knowledge. This problem extends into ecology. If one turns to ecological scientific journals, the inevitable findings will be particularistic, minutely specified research working off assumptions of ceteris paribus, that is, all things being equal. Through an assumption of ceteris paribus, isolation of a part and its changes can be subjected to reproducible tests of causation from independent variables. Tests purposefully isolate influences on the object to see which one can explain more, accurately. Precision is gained from mechanism, but context and the meaning and importance of constitutive relationships are lost.

One of the things we have learned (hopefully) in social science is that one method should not dominate and undermine other equally valid approaches, since the voice of research starts with its methodology. Relying only on mechanistic science reduces the diversity of voices and analysis that will be essential for creative and innovative thinking crucial for complex problem solving. Further, reading about the effects of one microinfluence in a microregion on a particular species needs to be balanced with interpretations of the "big picture" that these studies create. I acknowledge here that choosing holisms blinds me to the specifics of a part—in this case, the dynamics of a single country. I also acknowledge that looking at these parts is important work, and I am glad others do this work; but this effort is one which, for better or worse, is hoisted on the mast, looking at the horizon, not at the waves on the bow. This too is of value.

I will now discuss some epistemological frames for globalizing changes in the World Ocean. This thought experiment begins with complex systems theory.

Complex Systems Theory

Unlike Newtonian mechanism, complex systems theory is a nonlinear worldview. One cause can have multiple effects. Small forces can trigger large changes. Cause and effect live behind a thick and perplexing veil.

Complex systems theory sees connections and relationships as the most important factor in how a complex system operates. Parts are rarely, if ever, adequately described and understood in isolation. Rather, it is the picture of order that is important, but which occurs as a matter of chaotic localized action and interaction. Samir Rihani (2002), in his groundbreaking book on global development, writes, "Complex and Complexity refer to certain systems that have large numbers of internal elements that interact locally to produce stable, but evolving, global patterns" (6). This is not, however, simply a magnification of the smaller parts into a larger structure; such a magnification would amount to a linear understanding. Instead, complex systems theory sees systems as dynamic and "emergent" so that there is no deterministic view of cause and effect; same and similar conditions may provide different outcomes at different times. Further, Rihani explains that

> Despite the frenetic internal activity, outwardly the system seems to be unchanging. Occasionally, however, minor variations can trigger a major shift into an altogether different attractor [that which lends the global pattern] that presents a different global pattern, but there is no way of knowing which initial perturbation will shunt the system into a new pattern. (8)

The pattern is determined both by the number of parts in the system and the number of systematic elements each part controls, or each part's connectivity. Stronger, more stable systems become more complex over time in this fashion (Gell-Mann 1994).

For better or worse, the outcome of this system is not predictable: it is not linear, and changes in the system can come from minor alterations in the behavior of the parts, depending on the limits imposed by the rules of interaction. Too much complexity and connectivity, and the system can become overheated and end its connections (that is, in death); conversely, if the system oversimplifies and finds itself in absolute equilibrium, it will also die. Consequently, within complex systems theory, there is a preference for "ordered complexity" where the rules of interaction fall between equilibrium and chaos, allowing for open-ended transitions of the system.

Systems theory argues that, as impossible as it might be, (at least some) scientific attention should be focused on this big, global picture. Understanding these larger systems makes their functions and meaning clearer. It is as if mechanistic science has given researchers a magnifying glass when they really needed a telescope. Harrison (2000) further describes this theory through five main properties:

> First, complex adaptive systems have emergent properties. Second, they adapt to changes in their environment. Third, they are dynamically balanced between chaos and order, like life. Fourth, they draw energy from their surroundings to maintain their internal order. Fifth, complex adaptive systems are indeterminate and unpredictable. (104)

The World Ocean is a complex system. Its parts connect in dynamic and constitutive ways, changing in nonlinear fashion. Biodiversity connections and temperature exchanges with the atmosphere exhibit this nonlinear relationship where small changes in either of these two conditions can produce large changes elsewhere. What does it mean to read the World Ocean through this theory, and what can this theory tell us about sustainability? Are complex systems globalizing in a way that is a dead end? Some elements of marine biology, such as knowledge about the loss of coral reefs and collapsing/declining fish stocks, indicate that there is a diminishing diversity. Complex systems theory warns against such trends because these changes can promote systemwide changes toward total death of the system.

This theory can help us understand when something is global and when we can place a local point in the World Ocean within the larger system and see how that point is informed by the whole and helps to create it. Unfortunately, systems theory tends to see everything as part of a global whole; therefore, in order to make this a useful approach, I will argue that systems theory aids in making a global epistemology by looking for the structure into which parts play a role. If we can trace this function, we can argue that element is global. Visually, systems theory sees points connected to a matrix of other points so that each point in the matrix affects the conditions of the other elements to some degree, and the conditions of other elements affect that of the single point.

In addition to providing a way to conceptualize the meaning of connections in nature, complex systems theory offers a serious possibility of picturing human elements in the World Ocean. Ecologist and philosopher Gregory Bateson (1979) lays the groundwork for this thinking in *Mind and Nature*. Bateson argues that the mind and human behavior operate under the same parameters of nature, and that the "pattern that connects," the metapattern that other patterns are built upon and limited by, connects human mind and nature as they evolve together.

Hermeneutics—Reading Nature

Hermeneutics is not a method (Rogers 1996). It is an approach to knowledge that is aimed at fundamental, interpretive understanding. As Gadamer (1993) writes in *Truth and Method*, a person achieves this fundamental understanding through the "things in themselves," or what Heidegger describes as *dasein*. "For the interpreter to let himself be guided by the things themselves is obviously not a matter of a single, 'conscientious' decision, but is the 'the first, last and constant task.' . . . *The hermeneutical task becomes of itself a questioning of things* and is always in part so defined" (267–69; emphasis in original). Interpretation replaces observation through the reading of texts, as hermeneutic tradition began doing, and is done through a consistent criticism of one's own prejudice, which includes "fore-meanings." Fore-meanings are the biases we start projecting upon that which is to be interpreted. This provides a measure of insurance against capricious and arbitrary interpretation, but acknowledges that it is impossible (and undesirable) to eliminate our own standpoint and historical understandings as these provide a context to the whole of being. The importance and role of the oceans in people's lives, and their use of the ocean—their image of what the ocean is—can all be ways in which the ocean is "read." Geographers Adalberto Vallega and Stefano Belfiore (2002) have done work that is compatible with a hermeneutic approach. Vallega argues that knowledge about the ocean needs to be seen with reflection toward epistemological bias, but integrated holistically for more rounded understanding, which he describes as "post-modern." However, this postmodernism is one which is skeptical of neutrality, but not the project of building knowledge itself—much like the task of hermeneutics.

Gadamer (1993) describes language as the record which we are forced to reconcile with contemporary meaning:

> We regard our task as deriving our understanding of the text from the linguistic usage of the time or of the author. The question is of course, how this general requirement can be fulfilled. Especially in the field of semantics we are confronted with the problem that our own use of language is unconscious. How do we discover that there is a difference between our own customary usage and that of the text? (267–68)

Gadamer offers the suggestion that we know when the usage lacks meaning for the reader. I propose that this notion of understanding can bring new meaning and understanding of nature—perhaps a "reading" of nature. Nature, like language, is somewhat unconscious in our world—we are surrounded by nature in that we never leave nature just as we never abandon language.

"Reading" nature may produce (or it may not) a similar understanding of the ocean. We end up interpreting nature one way or another through projections of humans' own place in nature (e.g., in dominion over, in equity with, or below). These projections have historically determined the use of natural resources (Merchant 1980). Consequently, in making hermeneutics a theory of environmental understanding, I am really identifying the interpretation that is currently done and placing it in a deliberative and critical context for purposeful understanding of the ocean. This approach has important potential in environmental studies because it provides a way in which understanding "the thing in itself" helps us to achieve a global, though situated and contextual, understanding of, in this case, the World Ocean. Transformations of ocean ecology, while they may be disconnected from an objective human truth, can be read as *dasein* for meaning or even as the foundation of a great deal of meaning itself. In other words, if we lose coral reefs on a global scale, we lose meaning within the contemporary human experience—we are diminished in our potential and actual existence. The meaning of this loss shifts, but is not a simulacrum, or simulation that continually is rearticulated beyond recognition. The loss of a reef is genuine; where there was life there is now emptiness and attempts at ocean life to adapt to this absence.

In using this approach, it is critical to be conscious of our own "fore-meanings" and to concentrate on the World Ocean as something which has a life of its own that we can partially understand through comparison with our own dynamic concepts of what the World Ocean is like. Since context provides the fore-meanings, there is a danger of parochialism and conventionalism which may simply justify the current, dominant moral demands from the reader's cultural setting.

I do not propose that I, in this case a reader, have successfully escaped convention or relativism of my place and time (United States, environmental academia, rural family history, English language, etc.). But as Dan Sabia (2003) describes, this reading comes as an immanent, or internal and secondary, level of critique. The benchmark for this critique is sustainability, as set out and defined by Elisabeth Mann Borgese (1998). Borgese's notion of sustainability provides a compelling vision, which few societies, and surely globalization processes, fail to meet: ecological health comes together with social nonviolence, egalitarian governance, and material equity and subsistence. This provides a place to compare from the internal realm the situated "facts" of ocean changes as they relate to Borgese's idea of sustainability. Understanding how far away we fall from the "Borgese Test" will offer some meaning to our current ocean politics and give a critique of how to strive for these ideals. Since these are ideals, we will probably never achieve them fully, but then that never stopped societies from trying to achieve other ideals like freedom, security, and other aspects of the good life.

Critical Theory: Deconstructing the World and Putting it Back Together Again?

Critical theory is more interested in deconstructing meaning, power, and connections than it is in building them; nonetheless, I describe here how critical theory might be used to understand global phenomena as well as some thoughts about global "nature" from a critical theory perspective. While "critical theory" has been understood by the so-called Frankfurt School of German Western Marxism through the writings of Horkheimer and Adorno, Habermas, Marcuse, and Lukács, I refer to critical theory more as a systematic critique of modernity. Escobar (2003) clarifies:

> I understand modernity as a particular form of social organisation that emerged with the Conquest of America and that crystallised initially in Northwestern Europe in the eighteenth century. Socially, modernity is characterized by institutions such as the nation-state and the bureaucratisation of daily life based on expert knowledge; culturally, by orientations such as the belief in continual progress, the rationalisation of culture, and the principles of individuation and universalisation; and economically, by its links to various forms of capitalism, including state socialism as a form of modernity. (158)

Modernity has been the foundation for "development" politics from the beginning of the first wave of globalization. Escobar sees development and modernity as spatial-cultural projects that require continuous conquest of territories, people, and ecology to sustain it. Further, this project is now carried through the modern global capitalist economy, led by an imperial United States, "which seems more inhumane than ever," bent on exploitation particularly of the Third World (Escobar 2004, 208). As such, counterhegemonic movements which are locally based but transnational in effort are important for the preservation of people and ecology slated for violent displacement. This last point is one reason why, to preserve the World Ocean, I look to regional civil-society efforts that may be transnational and even globalizing.

Instrumental reason and scientific justifications for the manipulation of a mechanistic nature are compatible with the "development" of natural resources, which requires labor and its exploitation. This is not without political impact. Allan Schnaiberg (1980) describes the concentration of power that comes from ever-growing intensified withdrawals from and additions to the natural world as a result of intensifying and expanding capital—described as the "treadmills of production."

This cycle then expands outward, with global extensity. However, science, as the systematic pursuit of knowledge, cannot be blamed en masse. Specifically, Enlightenment science which is science based in modernity that pursues and promotes single notions of Truth and a pure objectivity through a separation of object/subject justifies the global exploitation of nature (Marcuse 1964); it is likely that current interdisciplinary science does not allow for exploitation in the same way because power in knowledge is more dispersed. Also, work in conservation biology, marine sciences, atmospheric sciences, and other important areas provides several reference points that legitimize resistance against expanding imperialism through the impacts of this economic expansion upon human ecology and vice versa. This assumes that interdisciplinary knowledge claims work against the concentration of power because authority is negotiated across epistemological commitments, instead of being self-reproducing within them (Daly and Cobb 1989).

I take several points from critical theory about globalizing changes. One is that the protection of human social diversity, and in particular the differences provided by the subaltern (antimodern, oppressed resistant groups), is important. Also, that knowledge serves certain interests; and scientific knowledge especially does so, given its credibility, which ironically comes from the improper assumption that it is objective and without ideology. Caveats and suspicions about science need to be kept alive so that the power in knowledge is not concentrated in any one purpose, interest, or part of the world. This is a theoretical reason to remain skeptical about, for example, fishery knowledge or methods that come from "globalizing" nations and attempts to replace other fishery knowledge, such as traditional artisanal fisher knowledge. It is also a reason to have more faith in the reverse because the orientation of power through modernity is organized against artisanal fishers. Finally, that agents of modernity seek to convert natural resources on large scales also homogenizes cultural values toward instrumental reason since instrumental reason pacifies resistance in the name of Mother Earth or other noninstrumental relationships with the natural world (Ridgeway 1996).

Critical theory provides a framework for viewing specific structural economic conditions and pressures through its neo-Marxist founding in conjunction with its critique of Western science and hegemonic culture. Consequently, critical theory is a balance against romanticizing the human "we" (even though I believe it is still a reasonable orientation) by looking at the power found in structures created to promote a globalizing interest. As Held and others (1999) note, all globalization studies need to confront modernity in some way; using critical theory as a theoretical framework is my way to do this.

Regional Orientation and Methodology

In this section, I describe why I have chosen to view ocean sustainability through regions. Regions are the level of analysis in this work; this may seem like an odd and irreconcilable starting point for looking at the "whole," but I believe there are good reasons to organize the study in this way.

I have chosen the Caribbean Basin, the South Pacific, and Southeast Asia as the places to look for globalizing and globalized conditions. One reason for this is that globalization processes, which are at the far end of the local, national, and regional continuum, need to go "across regions and continents" (Held et al. 1999). If there is a stretching of one ecological or social change across regions, this change is beyond the regional level and is globalized. The degree of globalization is subject to the minimalist and maximal scope described in chapter 1. This is the basis for the definition by Held and others of globalization, which includes the "transcontinental or interregional flows and networks of activity, interaction, and the exercise of power" (16). By looking at one region, we cannot say as much about globalization as if we compare activity between regions.

Therefore, my method is to compare the regions to their own history, to one another, and to what we know about the World Ocean in these regions, specifically regarding fisheries, coral reefs, and climate change. Each regional description starts with a historical sketch of experiences with globalization, then moves to a regional description and analysis of socioecological conditions.

Regions as the Level of Analysis

I now discuss in detail why I have chosen regions as the level of analysis. Regions are ecologically, politically, and economically salient, and I have chosen each of the regions in question based on these factors, using a "most similar" approach. The chosen regions are similar in that they have a high degree of dependence on the ocean through their large number of islands, and they have similar ecological facets given their coral reefs. Islands are important because they are defined by the ocean around them; this attribute makes them unique ocean spaces. This renders them distinct polities, and distinct ecologies, where high levels of endemism (species unique to an area) are regularly found. Thus, the high number of islands and island states found in each region was a main reason for choosing them. Also, these regions are similar politically, in that there is a degree of regional governance, organization, and identity in each.

Despite the preference in the field of international studies to focus on the nation-state, countries are not useful enough in looking at the whole. Looking at the nation-state places at the local level undue attention which will distract from the larger structural connections that I am trying to understand; distance from this frenetic difference will provide this ability.

Moreover, because the World Ocean is a regional and global common pool resource, nation-states are very often politically inadequate to deal with ocean problems on their own, but they may be able to deal with these problems in conjunction with other states at the regional level. The standing of the World

Ocean as a common-pool resource is disputable. Ostrom and Field (1999) delineate common-pool resources as areas where exclusion of beneficiaries is costly and exploitation by one user depletes the resource for others. The World Ocean is mixed on these terms. Access to seas, oceans, and the larger World Ocean is being limited more and more.

Keohane and Ostrom (1995) argue that even in places where the ocean is open, it (along with the atmosphere and biosphere) is not strictly a public good because of crowding where "the fisher who catches too many cod or the herder who grazes too many sheep in a pasture depletes the resource for others" (4). However, there are biophysical aspects and functions of the ocean that are nonexcludable and exhaustible, such as the depletion of coral reef through atmospheric forcing, or fishing on the high seas. On these counts the nation-state is not able to adequately manage World Ocean issues alone. Beyond these technical aspects, I also use the "common pool" term normatively, as did Borgese, in that the ocean is a world resource that is essential to everyone's health and well-being generally and that it should be used so as not to degrade these benefits for the world's people, plants, and animals.

Nonetheless, the nation-state is not invisible in regional and global study since nation-states can have a heavy impact on these larger arenas. In fact, regions are defined as "a group of countries with a more or less explicitly shared political project" (Hettne 1999, 1). Countries, however, are just one among many other agents within the region that make the region what it is, and this kind of effort is among other post-Westphalian work that looks beyond the nation-state to describe world politics.

At the other end of the spectrum in detail, the "whole earth" is not appropriate for a study such as this because there would be no possibility of variation. The areas of concern—ecology and politics of the World Ocean—would all look the same if we took the level of analysis to be the "Earth ocean," unless other planetary oceans were included. Therefore, regions provide a way to see globalizing processes at work, while allowing us to describe differentiation in the structure of such globalization at the same time.

Further, regions are increasingly recognized as important areas of governance for sustainability. The region is an appropriate level of analysis because ocean politics are being shaped more and more on the regional level *as a result of* globalizing problems. Hannah Cole (2003) writes that during the latter half of the century (when the Law of the Sea began its establishment and notions like the "common heritage" were developed), there began a shift toward the global in political organization to deal with several levels of increasing interdependence. Moreover, Cole argues that there is a continued disjuncture between the authority states claim and the structures in which they increasingly find themselves; as such, she argues that there is a transformation to other levels of governance such as regional and global ones.

This transformation of governance can be seen in relation to many oceanic areas of importance such as fisheries, wildlife management, atmospheric pollution and climate change problems, shipping, and other areas of interest which can not be wholly dealt with at the country level. This is one of the purposes behind the United Nations Environmental Programme's "Regional Seas" framework which has been in place since the 1970s. However, beyond the regional seas agreements which focus on specific coastal-management issues, this book looks to regions that have formed as a result of a regionwide identity and have created some kind of regional charter or treaty for international regionwide governance.

Analysis within the Region

I look at several structural conditions, divided roughly by history of globalization, regional oceanic changes, and regional political economy and institutions. Each regional history accounts for that region's experience with globalization. Typically, early globalization experiences are with the European and Iberian imperial efforts in one way or another starting in or around 1500. This history with globalization has had varying impact. For example, Caribbean experiences with colonization provide a common history that in turn provides a starting point for a regional identity. However, in the South Pacific, this history started later and was more diffuse.

Regional oceanic changes are organized around the three primary conditions I believe will have the most promise demonstrating global changes: fishery changes, reef loss, and climate change results found in rising sea levels and potentially changing currents. Each regional analysis contains an empirical accounting of these oceanic changes. Chapter 3 provides a background for these oceanic changes found more generally around the world; consequently, the regional conditions can be compared to these global trends.

I then describe the regional sociopolitical structure. Regional political-economic analysis includes an accounting of poverty in the area, in addition to a summary of violence at the level of "armed conflict." I count armed conflict using data from Eriksson, Wallensteen, and Sollenberg (2003) which is widely accepted in the literature on political violence. Armed conflict outside of their accounting will be, with some exceptions, invisible to me. This information, in addition to military expenditures, is part of the regional picture because, as Borgese described, nonviolence must be a part of sustainability. The absence or low level of armed conflict implies only a minimum measure of nonviolence, however, and should not be read as more than what it actually represents.

I also describe the degree of foreign direct investment (FDI), considering it a primary economic data point. FDI is capital invested into distant areas for production, and is a common metric of economic extensity and intensity (where and how much) of globalization. FDI is a way to manage distancing, discussed above. FDI can manage distancing from above by keeping the FDI mobile. Some of this capital, for example, is converted into fishing effort or coastal development, though the actual links to these endeavors are not very transparent and therefore must be presumed to some degree. If fish run out in one place, the FDI managers can move the boats to other locations without having to concern themselves about the initial stock. The trade institutionalized in FDI also allows for recurrent unequal trade in ecological energy, as discussed above. This trade would be managed through the particular kind of investment made—for example, FDI going into the United States is much more focused on higher entropic and more economically valuable goods and services than FDI is in Kiribati, where the focus is on lower entropic raw tuna catch.

As much as possible, I provide an analysis of capital flows specifically found in the fishing industry and coastal development. However, this capital is not always clearly delineated in its origin, and the literature on coastal development and knowledge production is not systematic, so these areas represent estimations gathered sporadically from the available literature. This comprises mostly secondary sources and primary documents available from public institutions.

Institutions are rules and roles, broadly speaking. The bounds of institutions are in "constitutions, international treaties, and culture" (Sharp 2005, 193). "When implemented, institutional structure provides a basis for the formation of commercial organization . . . the development and operation of markets . . . and the structure of nonmarket organizations" (Sharp 2005). Thus I look to very broad boundaries to estimate the process by which two of the above institutions, international treaty-based organizations and culture, seem to be moving.

I look at the basis for regional international organizations through the organization's history and its context for knowledge production to see if it is locally controlled and helpful in generating sustainability (for example, the role of indigenous knowledge is discussed where possible), and the regional organization's rhetorical framing and commitment to sustainability. I look at culture through the structure of social hierarchy in the region, illustrated in the ability of NGOs to act on a regional or global basis for outreach. Hierarchy is also implied by the degree of poverty and violence found in the region. I also describe the regional organization's orientation as evidenced in the organization's history and rhetoric.

The descriptions of the organizations are severely limited by space, and I admittedly paint with an overly broad brush, but there should be enough in these notes to see broad trends and form future research questions.

Findings are then interpreted through complex systems theory, hermeneutics, and critical theory. If the study is successful, I will have demonstrated several ways to see the World Ocean globally; I will have adequately explained the most urgent threats to ocean sustainability; and I will have convinced the reader that stemming these threats is inherent to a just and secure world.

Limits to Regionalism

One immediate limit is the practical ability to cover important details at the country or more local level. A better understanding of these regions would have been made if I had traveled to each area—even each country—and developed a deep understanding of the ecologies, cultures, and politics everywhere, but this was not possible. Instead, my descriptions of the regions are impressionistic macrosketches and use mostly peer-reviewed secondary information to build the cases.

In addition, it is worth noting that information is almost universally organized through, and monitored, controlled, and manipulated by the State. One clear example of this is the manipulation of marine fish catch numbers by the Chinese government: these numbers are understood to be grossly overestimated. A more subtle form of control is the lack of information on social conditions like poverty in the South Pacific, though this may be more of a symptom of a lack of capacity of or interest by the World Bank. Thus, most of the information used in this book is in some way compromised by this limit, which is part of a nationalist ideology.

Nationalism prefers distinctive sovereignty at the local level and perhaps even fragmentation and anarchy within the global system of states, as Waltz has argued (1959). In contrast, information gathered at the regional level is influenced by interests with a regional concern, and at this juncture this concern is admittedly less influential than nationalism. Another limit is that the variation at the state and locality levels will mostly be lost on this project, as was discussed above. I should state at the outset that I do not assume that any of these regions are homogenous in culture, politics, or other interests; as noted within, each area contains fantastic diversity that I value and mourn to see lost. I do assume that at the regional level there is a strong enough set of similarities compared to other regions that make this an interesting level of analysis for study. I appreciate the warning of critical theorist Greg Fry (1997) in discussing the South Pacific:

To presume the existence of something called "the South Pacific" need not necessarily lead to stereotyping of the peoples and societies within it. The extraordinary diversity—of cultural and linguistic forms, of resource endowments and geographical features, and of colonial experience—has been, after all, a major attraction for many of those calling themselves Pacific scholars. In such an approach, the region becomes a handy comparative frame as long as the basis of its construction is remembered. (Infotrac online)

Conclusion

There are many parts and pieces coming together to keep track of in this multifaceted set of theories and views. I believe this complexity is necessary in a mental experiment such as this study. No one has a God's-eye view of the world to see what is *really* happening. Or, if anyone claims to, we typically should not believe him or her. This means we are on our own to try to put things together from our inherently small space starting from our own homes, offices, libraries, conferences, correspondence, travel, and computers. Even discussions that are meant to confirm or disprove our assumptions and theories of holistic views will always lack a great deal of certainty. We are forcibly put into a place where conjecture, science, and philosophy uncomfortably meet. This tension is why I decided to start this scouting mission—to look for ways in which connections and glimpses of the whole World Ocean can be made—but the endeavor has not been easy. For example, physical science peer-reviewed journals rarely are in the business of presenting overviews of issues for non-physical scientists like me, and these overviews are not overabundant in the literature of the social sciences either. This means that a great deal of piecing together is involved.

By using three different epistemological lenses to view social- and physical-science information, I hope to provide three different avenues to seeing global ocean changes which I suspect are the result of the globalization of human impact from concerted human efforts, particular economic projects, and structures. I believe a particularly important outcome of this exercise will be the areas in which these frameworks can integrate with one another in such a way to include the best of each view. Complex systems theory can bring a sense of connectivity and complexity that seems to describe social and biophysical systems. That is, human systems can be described as having deep interconnections which constitute some of their ontological reality which makes them extraordinarily complex. However, complex systems theory lacks a way to interpret this complexity, and therefore seems to lack meaning. Hermeneutics can provide this meaning by attempting to subjectively read holistically into ocean phenomena described by complex relationships, and attempt to provide some sense of what these phenomena mean together. Critical theory, in addition to providing a specific skepticism for Western scientific and political economic structures, can ground the reading and descriptions in politically astute caveats that warn against overlooking detrimental conditions that may come with a global view itself.

Together, it seems plausible to build these connections and see how these frameworks can be integrated into a global epistemology that recognizes and uses science, and sees relationships as constitutive, but is not uncritical of its own position. From here, I take these lessons to look to the social- and physical-science literature for describing the World Ocean, reading into their connections as a way to understand these descriptions, and to self-criticize the way in which this reading relates to structures of power and knowledge.

Tenuous as they are, these efforts must be made. Oceanic science and interpretation need to be bridged on the global level in the attempt to understand anthropogenic causes of ecosystem changes that may have irreversible (at least on human time scales) effects when it comes to structural changes in the ocean. To do this, theory must be conceived, frameworks of analysis must be set in place, and mistakes must be risked.

Note

1. This chapter was presented to the International Studies Association in spring 2004, and I am grateful to the suggestions and discussions of Elizabeth DeSombre, Yannis Kinnas, Gabriela Kütting, and Radoslav Dimitrov. They are of course blameless for my errors.

Marine Political Ecology 3

IN THIS CHAPTER, I PROVIDE some basic necessary explanations of fisheries, coral reefs, and ocean temperature dynamics, as well as some important historical and political basics of ocean law and politics for readers who are not familiar with these issues. First, I provide a theoretical and geographic reasoning for thinking of the ocean environment as *the* World Ocean.

When one looks at the earth from space, it appears that the planet's major ocean basins are almost indistinguishable. Geographers identify different ocean(s) and seas, but the delineations between them often use latitude and longitude descriptions because land masses do not create absolute separations. The Pacific as a distinct basin is an imagined identification, not an essential or intrinsic one. All of the major ocean basins connect to one another. The Pacific Ocean connects with the Atlantic through the Bellingshausen and Weddell Seas and the Arctic Ocean; the Indian Ocean connects to the Pacific through the Timor, Tasman, South China, Andaman, and Banda Seas; the Atlantic connects with the Indian Ocean through the Antarctic/Southern Ocean. Sometimes these connections between oceanic areas are narrow and limited, as are the Straits of Gibraltar, which connect the Mediterranean to the North Atlantic. This geographic connection has allowed each wave of globalization that has relied on using the world's ocean as a highway. Trade can be seen physically going from one place to another, and this reinforces the use of critical theory that looks for resources and value to move away from the peripheral edge toward commercial centers.

At meeting points, ocean water, sediment, flora, and fauna are all normally exchanged. For example, Gage (2004) writes that "Basin confluence with the Atlantic, Indian and Pacific Oceans may have encouraged northwards dispersion of species from and into the deep Antarctic basins so that any regional identity is superficial" (1689). From the perspective of complexity, these flows of energy and material inform each other to create the outward face of what we know to be the ocean; without these substantive flows and relations between water, nutrients, organisms, and energy, the Earth would be a different kind of place entirely, and it is therefore enabled to be the biologically rich and stable planet it is because there is a global ocean with complex emergent properties. Hermeneutically, there is a connection between changes in the material and energy of the ocean and its meaning for humanity—in this perspective, the Earth is conceptually and materially defined by the fact that it has this global water body. Humanity's condition would be different if Earth's Ocean were something different—a little too salty and life changes dramatically, a little too warm or cold and life changes dramatically—all partly because the ocean is global. From each of these theories, then, we can conclude that if the ocean changes in any fundamental way, humanity is deeply affected.

One could reasonably object to the concept of a World Ocean. There are a few sites which are genuinely cut off from the global ocean, such as the Caspian Sea (and there is even deep disagreement as to whether this is a sea or a large lake; see, for example, Ascher and Mirovitskaya 2000). Another objection could come from ecological geography noting differentiated biomes based on differences in algae production and cycles (Longhurst 2001). Finally, it is not hard to imagine that each major ocean basin, connected through consistent economic connections between empires and colonies, can make up separate world systems in themselves (Cell 2004).

These are valuable perspectives, but none preclude the holistic view of seeing the ocean as a universally connected environment and social sphere. I choose to view the world's ocean environment as one interconnected fluid body. Imagine the counterfactual question, "What if the Pacific were entirely separated from the Atlantic?" Temperature stabilization would be very different, because the current system depends on the parts of the water column being interconnected. Fisheries would be isolated more. Colonial powers, tourists, and retail goods would not travel in the same way. Human history would be entirely rearranged. In sum, the fact that the Earth's ocean environment is physically and functionally connected is intrinsic to what this ocean is and means for the Earth and its inhabitants.

Introduction of Scientific and Political Notions of a Global Ocean

Some specific processes operate regularly through the whole. In this section, I set the context and reasoning for focusing on fisheries, coral reefs, and coastal development based on their relation to the universal World Ocean. I cannot overemphasize that this is necessarily an incomplete discussion of the terms, limited by my own background. Much more can be said about gas exchanges, chemical reactions, and the physics of how the water transfers salinity and nutrients, but these details are beyond the scope of this project and my own limited understanding.

I am particularly interested in three aspects of the World Ocean: fisheries, coral reefs, and the changes in the physical ocean water, including currents and sea level change. Each of these aspects represents global oceanic processes because they all affect the entire universe of the World Ocean, which affects the condition of the entire planet and offers a strong case for seeing the ocean in holistic terms.

Of the three aspects, the third is perhaps the most obviously global, whereas fisheries and coral reefs have very important local and regional dimensions that contribute to less obvious larger global ocean conditions.

Fisheries

Fisheries are measured as areas of the ocean where a specific marine species is taken for human use (e.g., FAO statistics are organized this way), though conceptually fisheries are a lattice of socioeconomic fishing efforts and networks that support fishing from the building of boats to the packaging of tuna for lunches of U.S. schoolchildren. Usually, the term *fisheries* refers to fish, but it also includes marine mammals such as whales and seals, crustaceans, and other types of creatures from the sea.

Fisheries can be very local, since some fish stay within a relatively small geographic area and typically do not leave these locations. However, there are at least three ways in which fisheries influence global oceanic conditions. First, as a major biotic group, fish in general are an important part of the world's gene pool. The second is that fish are an important protein source across the planet, particularly for Asians and the world's poor. Thus, fish are critical to the world food supply and the marine and human food chains. Finally, many marine species traverse great oceanic lengths; for instance, humpback whales and tuna do this, and they are global because they use transcontinental and transregional ocean habitat.

Changes to the world's fisheries indicate global change. Fisheries are changing structurally across regions, largely as a result of global fishing industry pressures. Large industrial fishing fleets hunt fish all over the world, enabled by refrigeration, sonar, and other advances that have changed the scope of fishing compared to historic and contemporary artisanal (small-scale noncommercial) fishers.[1] The large fishing fleets concentrate efforts on commercially valuable species which become more and more rare. After a species is commercially extinct, which means too rare and expensive to pursue systematically, commercial fleets move down the food chain to less commercially valuable species, thus stripping away biodiversity, complexity, and related food security. In the Black Sea, for example, populations of jellyfish have increased dramatically as their economically important competitors have been removed. This decline in the number of economically important species has caused a reduction in the number of main fisheries from twenty-six to five (Williams 1998).

Fish evolve against mortality pressures. Harvesting larger fish encourages evolutionary development away from larger-sized fish, instead favoring the development of smaller fish and losing the diversity and function of larger fish (Conover and Munch 2002). Notably, the fecundity (ability to reproduce abundantly) is partially dependent on fish length (Larkin 1978), meaning that impacts on fish size should also have an impact on fish's overall ability to reproduce effectively. Overfishing has globally extinguished vast amounts of fish species. According to the journal *Science*, historical research found that:

> Such ghosts represent a far more profound problem for ecological understanding and management than currently realized. Evidence from retrospective records strongly suggests that major structural and functional changes due to overfishing occurred worldwide in coastal marine ecosystems over many centuries. Severe overfishing drives species to ecological extinction because overfished populations no longer interact significantly with other species in the community. Overfishing and ecological extinction predate and precondition modern ecological investigations and the collapse of marine ecosystems in recent times, raising the possibility that many more marine ecosystems may be vulnerable to collapse in the near future. (Jackson et al. 2001)

The authors note that indigenous peoples pressed at least two species to extinction, but that most overfishing came as a result of later colonial and industrial pressures. Most importantly, the authors demonstrate that overfishing predated other human-induced marine problems such as eutrophication, hypoxia (loss of oxygen from the area), disease epidemics, toxic algae blooms, and other microorganism-caused problems which now occur in addition to overfishing.

In these fisheries, no more activity can occur without expecting declines in population, diversity, and resilience.[2] Current industrial efforts are wearing down trophic levels 0.05–0.10, where higher numbers indicate a higher level on the food chain from algae, every ten years *everywhere* in the ocean (Pauly et al. 2002). Some areas show drastic declines of predatory fish. The North Atlantic, for example, has experienced a decline of predatory biomass of 66 percent over the last fifty years (Christensen et al. 2003). The full effects of this process cannot be known, but they are not promising.

By most accounts, most of the world's fisheries are at or near their maximum levels, and some authors believe these numbers to be overly optimistic (Jackson et al. 2001). Figure 3.1 shows an impressive continuous increase in total fish catch around the world since 1950. However, looking at figure 3.2, we see that global marine catch in the wild (i.e., not aquaculture) has stayed steady or dropped in production levels from 1990 to 2001, while production pressures remain very high. Further, these figures take Chinese catch numbers as is, but the Chinese government, with the world's largest fishing fleet, is reported to have exaggerated its catch; this indicates that the global wild marine fish catch has actually been consistently declining since 1988 (Watson and Pauly 2001; Lu 1998).

In summary, the world's oceanic fisheries show signs of serious trouble. The political question follows: why? The answer is *still* usually found in two answers: common pool resource (CPR) governance problems and overcapitalization. CPRs are resources which can be depleted but are very hard or costly to limit access to or use of, so there is a problem of (a lack of) governance (Ostrom and Field 1999). In some areas, it is still common to find open access regimes, where there are no rules for access or use that are enforced. Both open access and a lack of governance on CPR fisheries are a legacy of mare liberum, which is product of the first wave of globalization. Mare liberum is discussed later in this chapter. Overcapitalization is a problem of investing too much money into fish exploita-

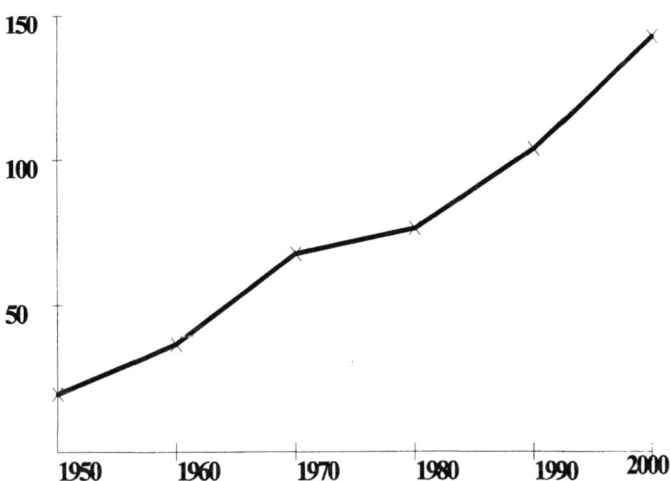

Figure 3.1. Global Wild and Aquaculture Fish Catch, 1950–2000
Source: FAO 2002.

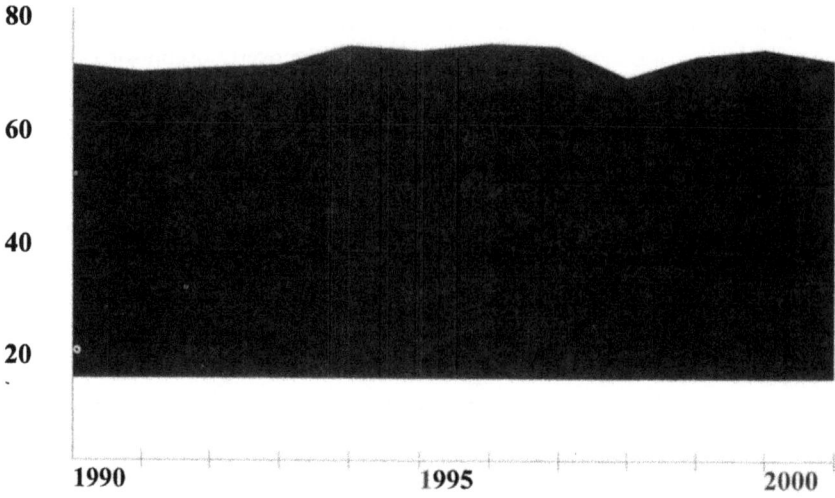

Figure 3.2. Global Wild Marine Fish Catch, 1990–2001
Source: FAO 2002.

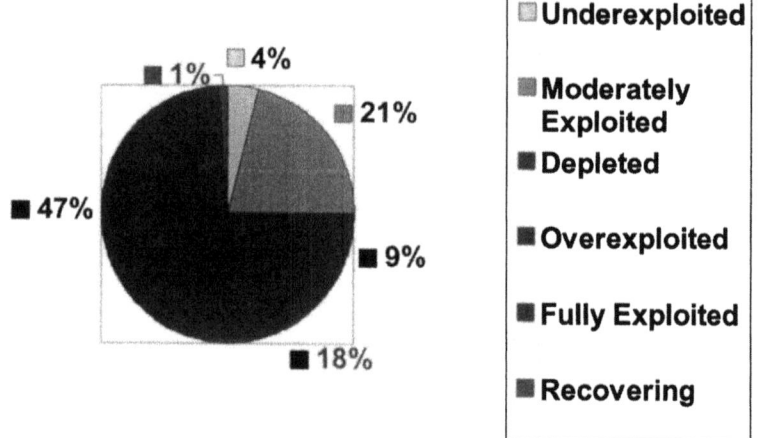

Figure 3.3. The Degree of Global Fishery Exploitation
Source: FAO 2002.

tion; this act creates competing economic and environmental interests. This issue will now be discussed in more detail.

Overcapitalization of Fisheries

The cause of fishery depletion is usually not very mysterious. Overfishing is the primary force behind the emptying of the ocean, though other forces are important. Despite international scientific and political acceptance that overcapitalization depletes fisheries, nationally and internationally subsidized fleets encourage overcapitalization. Every year, the fishing industry spends over $US120 billion for about $US70 billion in fish; the government subsidizes the rest (Prager and Earle 2000).

Overcapitalization is the problem of investing too much capital (money and equipment) into the fishery. The related term, *capacity*, includes the amount of time, efficiency of technology, and number of people engaged in fishing. Competition within a fishery encourages fishers to buy equipment that will net them— before their rivals—the greatest amount of the most valuable kind of fish. As the fishery suffers increasing problems in populations, size, weight, and diversity, economic pressures continue to drive fishers to take as much as they can. This problem has been internationally institutionalized through governments that use fish as an export commodity where growth in exports is needed for increased revenues and imports of other goods. Restricting the capital in a fishery after it is invested can mean temporary job loss. As the discussion below makes clear, fisheries used to be almost exclusively open access CPRs, but now coastal countries control the management of most fishing through exclusive economic zones. Nationalization of fisheries has increased management capacity but has not directed the incentive structure away from growth in landings, and some countries maintain de facto open-access fisheries that have little to no enforcement.

Without adequate restrictions, fisheries collapse and jobs and revenue are lost. Nonetheless, it is common for the fishing industry to pressure governments for high levels of catch; this practice follows state interests for exports. In the regional fishery boards that govern all U.S. fisheries, the fishing industry was overrepresented at an average of 49 percent of the membership from 1990 to 2001. Eighty percent of the 191 commercial fish stocks are fully exploited or overfished (Prager and Earle 2000).

Sport fishing interests make up another 33 percent, making fishing interests an overwhelming 82 percent of U.S. fishery board membership (Okey 2003). Typically, sport fishing organizations lobby for fishery conservation more than do commercial fishers, but nonfishing (e.g., non-fishing environmental organizations) interests are left with a small minority holding in governing U.S. areas. Also, the impact of sport fishing is now understood to be far greater than previously thought, given the concentration on top predators, continued open access of sport fishing regimes that do not limit users, larger numbers of sport fishers, and significant lethal and sublethal bycatch impacts. This leads Coleman and others (2004) to note that "Commercial and recreational fishing have similar demographic and ecological effects on fished populations. They truncate the size and age structures, reduce biomass, and alter community composition" (1959).

Investment of capital continues to be a problem. The long-run fishing capacity has increased enormously at a rate of 2.2 percent per year. The number of fishers in 1970 was about 12.2 million; now the amount is doubled to about 23 million with an additional 7.4 million farm fishers. Marine vessels numbered around 1.3 million decked and 2.8 million undecked (boats with decks have a larger capacity than, say, dugout canoes) vessels for a total of 4.1 million fishing vessels in 1998. The fastest growth occurred just before and during the 1980s phase of the third wave of globalization (FAO 2002), but has now flattened out.

Marine fishing rose exponentially and increased 300 percent directly after World War II. The volume of fish traded has increased even more: from $2.5–$2.8 billion in 1969–1971 to $35–$40 billion in 1990 (Garcia and Newton 1997). The world fish catch continues to grow (figure 3.1). Since the 1950s, total world fish catches have grown from around 20 million metric tons (mmt) to 130 mmt in 2000 (128 mmt in 2001) (FAO 2002). Meanwhile, 75 percent of the world's fisheries are fully exploited, depleted, recovering (1 percent), or overexploited (FAO 2002; Bruinsma 2003). Forty years ago, only 5 percent of the world's fisheries were fully exploited, overexploited, or depleted (Lubchenko 2003). This statistic matches the third wave of globalization of the 1950s, and the increased pace of capitalization corresponds to the latest phase of this third wave at the outset of the 1980s.

Thankfully, this trend is beginning to cool down, but the capacity of the fishing fleet is not managed for sustainability and will continue to wear down predator species. The amount of tonnage on the ocean fishing has leveled off since 1989, with small increases in world fleet size from 1990 to 1995. Note that marine catch from 1990 to 2001 (chart 3.2) is also relatively stable at this time. Interestingly, if capital were evenly distributed, the FAO estimates would reveal no overcapacity (FAO 1999). However, only 1 percent of China's fishing fleet is represented in the register, and it is assumed that the Chinese fleet represents about one third of the total (FAO 1999). Moreover, the world fleet is maldistributed, focused on the most lucrative predatory fish for commercial export economies. Consequently, the FAO admits the world fleet is 30 percent overinvested in major high-value stocks which make up 70 percent of the world catch, and that demersal (bottom-dwelling) fish populations in particular have been overexploited since the 1970s (FAO 1999).

These problems occur at least in part within the global political economy, or the world capitalist system which directs the flow of fish. Tables 3.1 and 3.2, and figures 3.4 and 3.5, show that so-called developed countries (I use the terms global North and South to refer to affluent and poor countries respectively) import the majority of the value in global fish trade. A great deal of the global fish trade is produced (caught) in the global South, and purchased by the global North where production is flat or declining. Furthermore, figure 3.5 shows intense capitalization of the global South's fishing fleet, which occurs so as to fill the orders of wealthy countries and orient the South to export-based economies, as proposed by neoliberal reforms in development and political economy. Much

Table 3.1. North-South Trade of Fish

Year	Developed Countries or Areas	Developing Countries or Areas	Developed Countries or Areas	Developing Countries or Areas
	Import Value (in constant thousands US$)		Export Quantity (in constant thousands US$)	
1976	7,656,728	1,185,339	5,396,753	2,524,039
1980	13,772,382	2,789,795	6,577,466	3,843,255
1985	16,277,404	3,213,294	8,531,166	5,342,197
1990	34,749,814	5,241,468	9,865,848	7,209,236
1995	48,193,504	8,873,716	11,561,225	10,836,781
2000	50,627,573	10,321,823	13,092,685	12,978,892

Source: FAO 2002.

Table 3.2. Global North-South Wild Marine Fish Catch

Year	Developed Countries or Areas (production quantity in metric tons)	Developing Countries or Areas (production quantity in metric tons)
1976	19,776,100	5,351,690
1980	20,090,101	7,322,875
1985	22,484,007	10,748,114
1990	22,510,228	14,441,812
1995	19,203,775	19,933,261
2000	18,972,832	21,842,474

Source: FAO 2002.

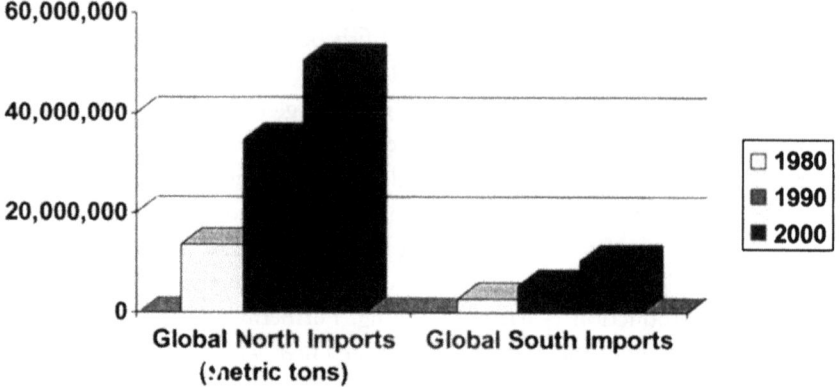

Figure 3.4. North-South Wild Marine Fish Consumption over 30 Years
Source: FAO 2002.

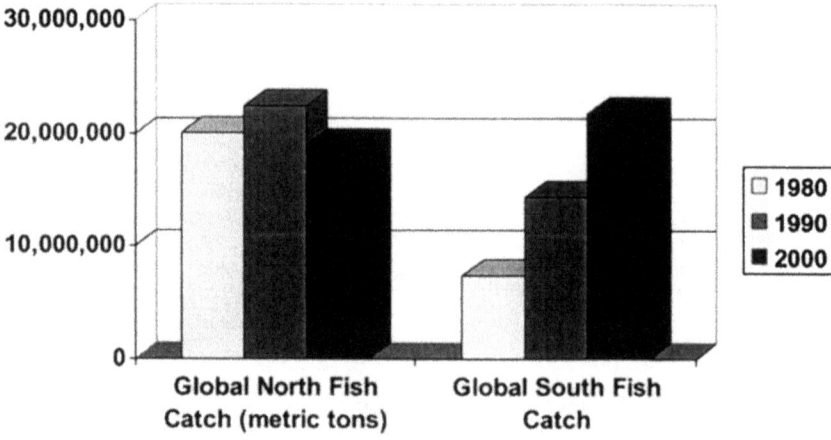

Figure 3.5. Scale of North-South Marine Fish Production
Source: FAO 2002.

of this capitalization came from development banks which lent the money to individual countries, which then had to participate in neoliberal policy changes that reduced public expenditures on social goods like health care, education, and even environmental protection. The resulting debt burden reproduces economic pressure for continued growth in fishing exports for the countries funding the

Coral Reefs

There are four primary types of reef: fringing, barrier, atoll, and platform. Each type of reef has varying functions, as well as varying flora and fauna, but for simplicity's sake I will treat them all within one category of "coral reefs." Coral reefs occur when corals and other organisms create calcium carbonate ($CaCO_3$) within a concentrated area. It often takes millennia for the calcium deposits to build up into reefs, but the time required for reef building varies for many reasons. The calcification building process depends on the density and extension of the calcium carbonate, which varies per geological zone and is partially dependent on ocean temperature. For example, Caribbean Sea coral seems to grow around 1.12–1.63 g cm^{-2} per year; this calcification rate has increased along with maximum sea surface temperatures, indicating response to warming sea temperatures (Carricart-Garnet 2004). Importantly, Carricart-Garnet points out, if this response is genetically fixed, coral may not have enough time to respond to further global warming. This failure to respond threatens the collapse or migration of these coral. If the calcification rate is not genetically fixed, coral may be able to adapt.

Reefs perform a variety of functions and services including the provision of coastal storm and surf protection, habitat for a wide variety of organisms, and even a land base for atoll (a horseshoe or circular reef around a central lagoon often very far from larger land areas) countries.

Coral reefs are at least a regional oceanic issue because they occur between and beyond localities, almost exclusively in tropical areas. However, coral reefs influence oceanic conditions globally through the numbers of fish and marine species they support. Thus, reefs are a major source of the world's gene pool. Anywhere from 600,000 to more than 9 million species are thought to use coral reefs for habitat (Knowlton 2001). About a third of the world's fish species live at one point in their lives within the network of global coral reefs, and about 10 percent of the world fish catch comes from these reefs (Moberg and Folke 1999).

Additionally, not only do fish need the coral reefs, the reefs need the fish. Overfishing within a coral reef area can harm reefs if herbivorous fish like parrotfish and surgeonfish, which keep down algae and allow for coral cover, are overexploited. For example, overfishing in Jamaica has shifted the coral cover of the reefs from 52 to 3 percent and algae from 4 to 92 percent (Holmlund and Hammer 1999).

Coral reefs are globally in danger; and there are no "pristine" or undisturbed coral reef systems left (Hughes et al. 2003). Currently, as many as 30 percent of the reefs in the world have already been severely damaged, and "Local successes at protecting coral reefs over the past 30 years have failed to reverse regional-scale declines, and global management of reefs must undergo a radical change in emphasis and implementation if it is to make a real difference" (Hughes et al. 2003, 929).

The usual suspects for this degradation are overfishing, inland pollution, global warming, predation by crown-of-thorns starfish, and coastal development. Reef fish are now part of a worldwide commercial chain starting, say, in the Philippines and ending up in Europe or North American fish tanks. Crown of thorns are a serious threat, for example, in Australia, where populations of the starfish break out and consume reefs; causes for their outbreak are suspected to be either high-nutrient land runoff from inland pollution (which can also kill reefs with too much sediment) or overfishing that removes the starfish's predators. Coastal development clears the reefs or nearby areas and increases pollutant and sediment loads on the reefs; these loads can degrade or kill the reefs.

The threat posed by "climate change, and regional scale bleaching of corals, considered dubious by many reef researchers only 10 to 20 years ago, *is now incontrovertible*" (Hughes et al. 2003, 929; emphasis added). Bleaching is the phenomenon of corals' expelling their zooxanthellae. Zooxanthellae are single-celled plants that live in the tissues of animals in coral. Bleached coral can survive, but many die; moreover, bleaching events can lead to mass coral death. As much as 16 percent of the world's coral was bleached and killed just in 1998, the largest bleaching event as of this writing.

Further, higher levels of the greenhouse gas, atmospheric carbon dioxide, will probably weaken the calcification process, and consequently slow reef building and/or weaken the skeletons of coral which form the structure of a coral reef (Kleypas et al. 1999). Thus, the atmospheric accumulation of gases emitted unequally is forcing a regional and global impact in tropical reef death and damage, which in turn will have a cascading effect on connected ecological species and functions. Therefore, coral reef decline and death appear to have several connections to the rest of the ocean, and it seems reasonable to consider coral reefs a global aspect of the world's ecology.

Climate Change and the World Ocean

The dynamic between the world climate and the World Ocean is complex, and is the subject of specific scientific inquiry by specialized scientists. Readers seeking more detailed analysis can find these scientists' work in such journals as *The Global Atmosphere and the Ocean System*. Here, I will present an oversimplified version of this knowledge, first about global warming in general and then as it relates to currents and sea level rise.

The World Ocean receives about 80 percent of earth's solar radiation, and absorbs about two-thirds of the earth's downward infrared radiation which results from greenhouse gasses in the atmosphere, for example, carbon dioxide and water vapor. Carbon dioxide and water vapor combine as the most important atmospheric greenhouse gasses which warm our planet; without them the earth would be about -19°C instead of the current 15°C, with water vapor about twice as influential as carbon dioxide.

However, a 30 percent increase in carbon dioxide above those found before the Industrial Revolution almost certainly has driven an increase in global temperatures. Some of this heat is released back into the atmosphere and some stays in the ocean for longer periods of time. The first 100 m of the water column absorbs nearly all of the heat, and most of the downward infrared radiation is absorbed by the top 1-cm layer. Even in the top 100 m, though, temperatures rarely change more than 10°C seasonally, and rarely more than 0.001°C daily (Mason 1993). Thus, the ocean, and the deep ocean in particular, changes temperature very slowly due to enormous thermal inertia stored as a result of long-term heat transfer. It takes a long time for heat to transfer from the top layers of the ocean to the lower layers, but once that heat is transferred, it stays there for a long time, traveling slowly through the global current system. Consequently, increased heat absorption in the oceans, particularly in the deeper ocean levels, will drive increased global temperatures well after carbon dioxide emissions are reduced. Since the 1950s, the heat content of the ocean has increased globally (IPCC 2001).

Mason (1993) writes, "Heat transfer by the eddies and the mechanisms of vertical heat transfer are not well understood. There is, however, no doubt but that the response time of the oceans, ranging from decades in the top 1km to centuries in the deep ocean, sets the pace for global climate change" (20). In addition, carbon itself is absorbed by the World Ocean; this phenomenon is called the "solubility pump" in the ocean because it relates to how soluble carbon is in the ocean. The solubility pump depends on ocean temperature, the degree to which water mixes vertically in the water column, and global circulation patterns (Geider 2001). The degree to which the ocean water itself is capable of absorbing this carbon is therefore essential for a global carbon sink (where waste goes), but the conditions which dictate this process are themselves changing. About half of the released carbon from 1800 to 1994 has been absorbed into the

World Ocean, raising the carbon dioxide level from its 400,000-year constant of 280 parts per million to about 380 parts per million, which slightly exceeds the atmospheric levels at the same times (Takashi 2004; Sabine et al. 2004). A doubling of carbon from 1800 levels is impending (Takashi 2004; Sabine et al. 2004).

This change in the ocean chemistry is expected to harm calcium-building organisms, for example, shellfish, some phytoplankton, and coral (Feely et al. 2004). Thus increased carbon from the atmosphere and the water is pressuring calcification in addition to the rising temperatures, causing bleaching. This means that reefs are threatened by global warming processes and results. A statement from over 100 oceanographers who have assessed carbon absorption in the ocean notes that the "effects are already occurring" and this change in chemistry "could have a significant negative effect . . . disrupting marine food webs" (Revkin 2004, D3).

The impacts of carbon on the ocean illustrate a key point for this book, that the World Ocean is a set of connected subsystems that impact not only the larger ocean, but all of life on the planet. For example, without the ocean's carbon absorption, the atmosphere might have double the carbon dioxide it now harbors. One change informs another, specifically, within climate changes that interact with a multitude of levels—from wind, wave activity, wildlife and flora interactions, and even the role of the Earth's rotational spin. I will now specifically limit my discussion, however, to currents and sea level changes, both of which are experiencing systematic, transregional alterations.

Currents

Currents move over the globe regulating temperature and marine production, such as that of phytoplankton (basic plant organisms that move only by floating), which produce huge amounts of oxygen. Deep thermohaline currents (defined by saline content and temperature) run at the bottom of the ocean at about 2°C and at 1 mm per second—taking about a thousand years to circulate the globe. This current keeps the earth's temperature stable, and there is evidence that the flow of these currents may change along with global warming, but ocean research of this type can be difficult and unsure (Wood et al. 1999). Nonetheless, the importance is clear: "changes in the formation of deep water masses at high latitudes in the North Atlantic and the Southern Ocean could lead to abrupt changes in the global ocean thermohaline circulation and a major rearrangement of global climate" (Grassl 2001, 7).

The oceanic currents revolve around the earth, maintaining its relatively stable temperature. Cold water from the poles circulates down to the equatorial regions, and the warm equatorial waters flow up to the poles in a "conveyor belt" of temperature regulation. The international group of over 2,000 scientists charged with studying climate change, the International Panel on Climate Change (IPCC), warns that this function may change as a result of global warming's raising the temperature of ocean waters:

> Most models show weakening of the ocean thermohaline circulation which leads to a reduction of the heat transport into high latitudes of the Northern Hemisphere. . . . The current projections using climate models do not exhibit a complete shut-down of the thermohaline circulation by 2100. Beyond 2100, the thermohaline circulation could completely, and possibly irreversibly, shut-down in either hemisphere if the change in radiative forcing is large enough and applied long enough. (16)

Further, as the conveyor belt of warming and cooling water is shuttled around, this water influences air temperature; this in turn influences weather and weather patterns. Over a long period of time, patterns of weather are categorized as climate. Through these flows, water, chemicals, gases, nutrients, flora, and fauna exchange with similar elements in different areas of the world, creating an ecological platform for globalizing these exchanges.

One of these gas exchanges occurs through phytoplankton. Phytoplankton are the ocean's primary source of food: they provide oxygen, and they do this in exchange for carbon dioxide. Prager and Earle (2000) explain:

> The smallest, most vulnerable—and yet quite possibly the most important—creatures of the sea are the floating marine plants and algae, the phytoplankton. The phytoplankton are at the base of the ocean's sun-driven food web (to differentiate from the chemical-driven web at deep-sea vents). Through photosynthesis, they produce organic matter from the sea's inorganic materials, this is often called *primary production*. (191)

So in addition to the solubility pump, the World Ocean provides a carbon sink through its phytoplankton, referred to as the "biological pump." Algae exchange about 48×10^{15} g of carbon per year compared to the approximately 56×10^{15} g of carbon exchanged terrestrially (Geider 2001). After the algae dies, it sinks to the bottom of the ocean, virtually locking away the carbon. This function is limited by available nutrients, and is not expected to be important in further gains for the biological pump as a sink (Geider 2001). However, "changes to the biology of the pump may be the most critical component of the oceanic responses to future changes in climate and CO_2," but the nature of this response is too difficult to predict right now (Geider 2001, 853).

This process accounts for a little less than half of the planet's oxygen and therein is the second most important and fundamental global function of the World Ocean (Day 1999). Since this oxygen is not distinct from other supplies of oxygen, every living creature that requires oxygen to live on Earth depends on the World Ocean to do so. A change in either the temperature regulation or the oxygen production function of the World Ocean would have, by definition, global consequences throughout the entirety of the planet without important exception.

Sea Level Rise

With increased global surface air temperatures, sea ice and ice sheets are melting. Sea levels rise as a result of melting ice sheets on land which drain into the ocean, and as a result of expanding water volume in the ocean, which is a result of warmer water temperatures. As the ocean currents influence the atmosphere and the combination of the two influences composite climatic conditions in an apparent warming trend, coastal shorelines advance toward land. This event also has detrimental impacts on small-island freshwater supplies, urban infrastructure, and human health.

Small island countries face severe threats of rising waters, since most are not more than 3–4 m above sea level. In addition, many small island countries are poor and use relatively few hydrocarbon resources compared to industrialized countries with a larger land base, as noted in tables 3.3 through 3.6, which depict carbon emissions by country.

Consequently, small island countries face a problem of equity. Large industrial countries enjoy the benefit of large-scale hydrocarbon use, but face smaller climate costs than do small islands, some of which may actually disappear as a result of large-scale hydrocarbon emissions like carbon dioxide. Since larger nations will not suffer the same costs, it will be hard for these politically weaker countries to change the emissions policies of larger, more powerful countries. The scale of this usage is demonstrated by figure 3.4, which shows that the United States outpollutes its closest peer, China, in total carbon emissions by double. In per capita emissions, however, China uses 0.6 metric tons per person per year compared to the United States' emissions of 5.4 metric tons per person per year.

The global average sea level has risen two millimeters a year since the mid-1800s and is one of the most evident signs of global climate changes, resulting in

Table 3.3. Top 25 Total Carbon Emission Countries in 2000

	Country/Territory/Commonwealth	Total Carbon Emissions* (in thousand metric tons)
1	United States of America	1,528,796
2	China (mainland)	761,586
3	Russian Federation	391,664
4	Japan	323,281
5	India	292,265
6	Germany	214,386
7	United Kingdom	154,979
8	Canada	118,957
9	Italy (including San Marino)	116,859
10	Republic of Korea	116,543
11	Mexico	115,713
12	Saudi Arabia	102,168
13	France (including Monaco)	98,917
14	Australia	94,094
15	Ukraine	93,551
16	South Africa	89,323
17	Islamic Republic of Iran	84,689
18	Brazil	83,930
19	Poland	82,245
20	Spain	77,220
21	Indonesia	73,572
22	Turkey	60,468
23	Taiwan	57,991
24	Thailand	54,216
25	Democratic People's Republic of Korea	51,544

Source: Adapted from the Carbon Dioxide Information Analysis Center at cdiac.esd.ornl.gov/trends/emis/top2000.cap by Marlen, Boden, and Adres.
*Carbon emissions from fossil-fuel burning, cement production, and gas flaring.

about 10 cm of higher sea levels and is expected to continue to rise 4 cm per decade assuming an increase of 0.3°C per decade. This 10 cm rise is thought to comprise 4 cm from thermal expansion of ocean water, 4 cm from mountain glacier melt, and 2–5 cm from increased melting of the Greenland ice sheet (Mason 1993). As the average surface temperature of the earth has increased over a half of a degree Celsius (1°F), the rate of sea level rise increased tenfold (Douglas, Kearney, and Leatherman 2001). Estimates vary, but the earth is expected to increase in temperature by 1.4°F to 5.8°C (2.5°F–10.4°F) (IPCC 2001). Glaciers and ice sheets are melting, and the polar regions are warming on average of 4.5°F to 5°F. The Greenland ice sheet alone is losing about 51 cubic *kilometers* of ice each year, which is sufficient to raise the sea level 0.13 mm/yr (Krabil et al. 2000). According to a 2004 study by more than 250 international scientists,

Table 3.4. Top 25 Per Capita Carbon Emissions in 2000

Country/Territory/Commonwealth	Carbon Emissions Per Capita (in metric tons)
1 Antarctic Fisheries*	61.12
2 U.S. Virgin Islands	29.91
3 Qatar	19.65
4 Netherland Antilles	12.61
5 Bahrain	7.7
6 Guam	7.17
7 United Arab Emirates	6.17
8 Kuwait	5.97
9 Trinidad and Tobago	5.58
10 United States of America	5.4
11 Luxembourg	5.31
12 Falkland Islands (Malvinas)	5.24
13 Aruba	5.2
14 Brunei (Darussalam)	5.08
15 Wake Island	5.02
16 Australia	4.91
17 Saudi Arabia	4.77
18 Singapore	3.9
19 Canada	3.87
20 Faeroe Islands	3.84
21 Palau	3.48
22 Estonia	3.19
23 Czech Republic	3.16
24 Nauru	3.07
25 Ireland	3.04
26 Norway	3.03

Source: Adapted from the Carbon Dioxide Information Analysis Center at cdiac.esd.ornl.gov/trends/emis/top2000.cap by Marlen, Boden, and Adres.
*The authors of the original data note that "the listing of the Antarctic Fisheries as the largest emitting nation/territory/commonwealth is incorrect due to a data omission for 2000. The largest source of emissions from the Antarctic Fisheries are from importing and use of diesel fuels. Much of the imported diesel fuel is consumed by other nations during international or 'bunkers' trading. The 2000 estimate of 'bunkers' diesel use was omitted from the UN Energy Statistics Database thus giving the false impression all imported diesel fuels were consumed by 'Antarctic Fisheries.'"

Arctic ice is half as thick as it was thirty years ago. Furthermore, warming in this region is accelerating faster than previously thought and at current rates may be free of ice by 2070 (Arctic Climate Impact Assessment 2004).

Increase in sea volume can continue for a very long time since nearly all fresh water found on the planet is frozen in ice sheets, glaciers, and ice caps. About 97 percent of the world's water is already in the ocean. However, of the remaining 3 percent (the only freshwater on the planet), 2.997 percent of the world's water is frozen in glaciers and in the polar ice caps. Currently, about one-third of the

Table 3.5. Bottom 25 Total Carbon Emission Countries in 2000

Country		Total Carbon Emissions (in thousand metric tons)
188	Gibraltar	59
189	Grenada	58
190	Equatorial Guinea	56
191	St. Vincent and the Grenadines	46
192	Solomon Islands	45
193	Cape Verde	38
194	Samoa	38
195	Nauru	37
196	Chad	34
197	Tonga	33
198	Dominica	28
199	St. Kitts-Nevis	28
200	Sao Tome and Principe	24
201	Comoros	22
202	Vanuatu	22
203	British Virgin Islands	16
204	St. Pierre and Miquelon	15
205	Montserrat	13
206	Falkland Islands (Malvinas)	10
207	Cook Islands	8
208	Kiribati	7
209	Wake Island	5
210	Saint Helena	3
211	Niue	1
212	Turks and Caicos Islands	0

Source: Adapted from the Carbon Dioxide Information Analysis Center at cdiac.esd.ornl.gov/trends/emis/top2000.cap by Marlen, Boden, and Adres.

yearly sea level rise worldwide is attributed to glaciers; this contribution has doubled over the last twenty years. From 1961 to 1976, about 10 percent, or 0.15 mm/yr, was attributed to glaciers, but during the years 1988–1998, that contribution increased to 0.41 mm/yr, or 27 percent of global sea level rise (Dyergurov 2003).

The transition of extremely large ice sheets into the ocean continues to make headlines. In 2002, a piece of the Antarctic ice sheet the size of Connecticut broke off (Revkin 2002). The largest ancient ice shelf in the Arctic, the Ward Hunt Ice Shelf, which has been at the northernmost part of Canada for at least 3,000 years broke up over 2000–2002, from the increase in temperatures in the Arctic over the last century. Ward Hunt was part of a more extensive shelf, Ellesmere Island, which has shrunk by over 90 percent over the same time (Revkin 2003).

Table 3.6. Bottom 25 Per Capita Carbon Emissions Countries in 2000

Country		Per Capita Carbon Emissions (in metric tons)
188	Sudan	0.05
189	Eritrea	0.05
190	Guinea	0.04
191	Nepal	0.04
192	Madagascar	0.04
193	Liberia	0.04
194	Sierra Leone	0.04
195	United Republic of Tanzania	0.03
196	Comoros	0.03
197	Niger	0.03
198	Burkina Faso	0.02
199	Lao People's Democratic Republic	0.02
200	Rwanda	0.02
201	Central African Republic	0.02
202	Uganda	0.02
203	Malawi	0.02
204	Mozambique	0.02
205	Zaire	0.01
206	Mali	0.01
207	Afghanistan	0.01
208	Cambodia	0.01
209	Burundi	0.01
210	Chad	0
211	Turks and Caicos Islands	0
212	Ethiopia	0

Source: Adapted from the Carbon Dioxide Information Analysis Center at cdiac.esd.ornl.gov/trends/emis/top 2000.cap by Marlen, Boden, and Adres.

Without ice sheet collapse, the sea level may rise as much as 1 meter in the next 100 years, putting small islands like Tuvalu in the South Pacific on alert that their country will disappear in about 50 years. The thousands of islands that constitute the Maldives nation are typically at 1–1.5 meters above sea level. A one-meter rise would essentially wipe out this country as well (IPCC 2001), though some research is more optimistic because the Maldives have survived previous sea levels this high and are in fact now seeing a sea level drop attributed to high levels of evaporation confined to the central Indian Ocean (Mörner, Tooley, and Possnert 2003). Also affected would be coastal dwelling people on continental landmasses, particularly those people who live at river deltas. Under current projections for sea level rise in the next 100 years, over 100 million people would be displaced in China, Bangladesh, Egypt, and Nigeria alone, not only due to a nearer shoreline, but by increased storm surges and erosion.

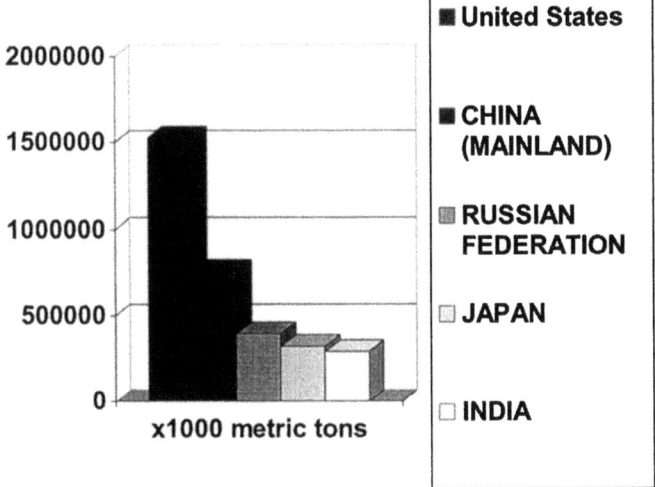

Figure 3.6. Scale of United States Carbon Emissions
Source: Adapted from the Carbon Dioxide Information Analysis Center at cdiac.esd.ornl.gov/trends/emis/top 2000.cap by Marlen, Boden, and Adres.

From Mare Liberum to the Common Heritage of Humankind

Since oceans are connected, vessels can move from one sea to the next without encountering many geophysical barriers. Some polities have the economic resources to establish, maintain, and build fleets which are capable of traversing these distances, and some are and historically have been at a relative disadvantage for doing this. Typically, colonial powers used their capability to send fleets of ships across large oceanic distances to extend the powers' reach, without which imperialism would have had an entirely different face. This ship-based power was used to acquire more material power in the form of natural resources for military and commercial uses. This history then allowed for material power to accumulate within these imperial forces, which then institutionalized many of these countries as centers of influence for decades to come. For example, in the period from 1503 to 1550, Spain used the capital of about fifty tons of gold removed from Caribbean Antilles to fund further regional colonization (Sued-Badillo 1992). Some countries did not maintain this power, but the current global North-South division is roughly organized between countries that were able to organize, expropriate, and accumulate foreign natural resources and labor through naval capabilities.

This difference became very pronounced between "distant water fishing nations" and "coastal nations" during the negotiations of the Law of the Sea, where distant water fishing nations were almost exclusively former colonial powers. As it turns out, this division established the foundation for international ocean law for hundreds of years.

Mare Liberum *and the Open Oceans*

In order to understand contemporary relations with the ocean, it is necessary to look back to the seventeenth century, when the course for instrumental human-ocean relationships was globally imposed over noninstrumental relationships found elsewhere (Jacques 2001).

In 1609, Hugo Grotius wrote one of the most important international legal doctrines regarding the oceans, *Mare Liberum*, which translates from Latin to "the freedom of the seas." Mare liberum states that the ocean, specifically the high seas, could not be owned by any person or nation. Consequently, the oceans were free for anyone to use and exploit without limits. Grotius was primarily concerned with the navigation of the oceans and keeping shipping lanes open, but the effect of his doctrine expanded to all uses, such as fishing and other industries. Mare liberum persisted in this form until the mid-twentieth century. Today the concept still has relevance but has been mitigated practically by increasingly monitored and regulated oceans, and even on the high seas there are formal limits to activity on the ocean. Nonetheless, the legacy of *mare liberum* governance is an assumption of negative freedom on the ocean, where if an activity is not explicitly prohibited it is permitted, particularly on the international high seas.

The territorial seas were the only exception to the *mare liberum* doctrine. Territorial seas were measured out three nautical miles (a nautical mile is 6,076 ft, slightly longer than the measurement of a terrestrial mile, 5,280 ft) from the low tide of a coastal nation.

Mare liberum freedom assumes that the ocean's resources are inexhaustible. Nations saw the ocean as unlimited in its vast stretches of space, unlimited in its marine life, and unlimited in its ability to take on new users and waste. Resources such as fisheries, or areas that are fished, were seen to be so plentiful that human endeavor could not hope to affect the numbers of fish available. Therefore, limits on fishing seemed ludicrous at best and economically damaging at worst. As reports of low fish catch by fisherman and some preliminary scientific studies in the 1800s came to challenge this assumption, the international community began to move toward protecting resources that were commercially important.

The oceans were understood primarily in international terms. At this time it was beginning to become obvious that what one country did on or to the ocean fundamentally affected what other countries could do with the ocean. Nonetheless, it was not until the late nineteenth century that the doctrine of the free seas was seriously questioned internationally, and it was not until 1958 that the free seas doctrine was limited in any way by multilateral international laws, and even these laws did not challenge it.

As a result of this colonial framing, the primary international view of the oceans has been that of relatively unrestricted, open use. Thus, most international ocean laws exist as exceptions to that freedom. Limiting or putting boundaries on this freedom was the primary task of the United Nations Conference on the Law of the Sea. There were three of these conferences, starting in 1958 with Law of the Sea-I and ending in 1982 with the closing of Law of the Sea-III. The premier concern for each of these conferences was how to manage the well-established freedom of the seas in light of new knowledge that the oceans were more fragile than previously thought.

Management of the oceans, which usually means limiting or encouraging human practices that affect the ocean, is fundamentally in conflict with the notion of the freedom of the seas. Without implementing these limits, overfishing, pollution, wanton development, and degradation occur, as they will with any open-pool resource that is devoid of rules of access or use (Hardin 1968; Buck 1998). This threat is not inevitable. Empirical social science research consistently shows that local and global resources in particular are amenable to nonhierarchical resource management through constraints derived by revised incentives (Keohane and Ostrom 1995). Constraints are being developed through regional and multilateral treaties, but some ocean scholars still believe that "the 'anything goes' ideology still prevails" (Van Dyke, Zaelke, and Hewison 1993, 3).

Significant limits to exploitation of ocean resources are in conflict with contemporary neoliberal politics; that is, neoliberal ocean politics are compatible with mare liberum. This implies that strict neoliberal policies, to the extent that they obstruct limiting human impacts, are also incompatible with long-term ocean sustainability since both provide incentives for overexploitation. Sustainable ocean management requires institutions to develop local, regional, and global incentives for limits to use of and access to ocean resources including fisheries, coral reef areas (e.g., limits to coastal development), and climate sinks. Since historical precedent and the current globalization of neoliberalism militate against this need, ocean sustainability has a steep slope to climb.

This is perhaps why the final conference of the Law of the Sea took so long to negotiate. Determining one kind of limit, for example, a large area of coastal control, would work in the favor of the coastal states that then would benefit from exploiting that larger area previously open to the more powerful countries.

In the end, the Law of the Sea created a favorable framework for poor coastal nations through the large 200-mile zone of control. (Sovereignty of coastal areas now extends out to 12 miles, but the coastal country manages the full 200 under normal circumstances, so in principle poor countries can exclude distant water-fishing nations from fishing local stocks.)

However, more importantly, the Law of the Sea set the framework for a sustainable ocean relationship. For example, ocean resources are to be used only for "peaceful purposes," and the Law of the Sea established the first natural resource-based scheme for global redistribution of wealth, called the Common Heritage.

The Common Heritage of Mankind is an idea that was introduced by the Maltese delegate, Arvid Pardo. The Common Heritage of Mankind prescribes that all minerals taken from the deep seabed within "the Area" belong to and should benefit all of humankind, not just the countries that have the ability to retrieve said minerals.

> Activities in the Area shall, as specifically provided for in this Part, be carried out for the benefit of mankind as a whole, irrespective of the geographical location of States, whether coastal or land-locked, and taking into particular consideration the interests and needs of developing States and of peoples who have not attained full independence or other self-governing status recognized by the United Nations . . . (XV) and other relevant General Assembly resolutions. (United Nations Law of the Sea, 2001 online)

The "Area" is that section of the seabed under the high seas; however, Borgese and Pardo both argued that most areas of the ocean, not just the current Area, should be the common heritage of all people, given the global nature of the World Ocean.

One of the resources on the deep seabed is in the form of manganese and polymetalic nodules that were discovered in the 1870s by the HMS *Challenger*. The nodules sit at the bottom of the ocean and resemble small, misshapen iron canonballs. The nodules still are not able to be mined in an economically profitable way, but delegates to the Law of the Sea, and Pardo and Borgese in particular, saw this resource as one that had not yet been monopolized but could one day be used to serve global peace and the uplift of the world's poor.

Pardo warned that without a common heritage, the "strong would get stronger, the rich richer, and among the rich themselves there would arise an increasing and insuperable differentiation between two or three and the remainder" (Juda 1996, 189). Since weaker marine states did not have the same capacity as strong marine powers to collect these minerals, Pardo wanted to make sure that the resources were not used to reproduce global inequity between rich and poor states.

As a result, industrialized countries balked at the Law of the Sea. In order to bring in dissenting countries, a further 1994 agreement was conducted specific to Part XI, the section in the Law of the Sea which describes deep-sea mining. The premise of this additional agreement was to allow mining of the Area by private industry in addition to the International Seabed Authority, the agency established under the Law of the Sea to manage the distribution of revenue from the mining. Originally, the common heritage intended mining of the Area to be solely an international public works effort. Some analysts say this extra agreement actually bargained away the major benefits for the world's poor since private industry would essentially be competing against the international governing efforts. Industrial nations, however, were not willing to totally replace profit with charity (Borgese 1998), and they can now lease rights to mine in the Area, mitigated by the International Seabed Authority, which is headquartered in Jamaica.

In conclusion, the struggle to govern the oceans was one of the first tasks of the first wave of globalization. Current institutional trends have made progress toward sustainability compared to the bleak emptiness of mare liberum, and important textual meaning for genuine sustainability has been imbued into the texts of several ocean documents like the Law of the Sea. Other promising examples are the related Fishery Stock Agreement of 1995, which allows regional authorities to clamp down on violators of fishery conservation for migratory species, like tuna, that cross political borders, and Chapter 17 in Agenda 21, which establishes baseline coastal conservation policies under precautionary assumptions. All of these treaties are relatively new for ocean politics, and to the extent that countries and regions adhere to them, the prospect for ocean sustainability is greatly enhanced. They all treat the ocean as a whole, apply to the entire ocean, and attempt to establish relationships with the World Ocean that acknowledge that "the oceans are the very foundation of human life" (United Nations Division of Ocean Affairs and the Law of the Sea website www.un.org/Depts/los/oceans _foundation.htm). This aspect should give us optimism about the processes necessary for sustainability within the World Ocean—but as I show through the rest of this book, this optimism would be entirely premature.

Notes

1. Artisanal fishers may sell their fish, but it is generally caught for local consumption, not national export.

2. I understand that the function and reality of "resilience" is disputed in the biological literature. From a human perspective, though, dynamic equilibrium of a fishery remains important. Dynamic equilibrium implies nonstructural changes to the fisheries so that we continually have new and replaced individuals but that the operation of the fishery itself continues to be relatively stable over time.

Sustainability in the South Pacific 4

THE SOUTH PACIFIC REGION is the most sustainable region analyzed in this book.[1] The region has many problems, its ecology faces serious threats, and it even boasts one of the most well-known cases of ecological collapse, Easter Island. However, the region has a strong legacy from indigenous institutions and law, and it is purposefully organized around environmental sustainability goals, the diffusion and regional creation of knowledge, and participatory regional governance. The region does not meet all the sustainability goals it has set for itself, and is up against some serious problems. Lobban and Schefter (1997) warn that, "Many of the environmental changes that occur on Pacific Islands are similar to those occurring elsewhere in the world, but the changes may be greater or swifter in these miniature worlds" (21). However, if sustainability is a process instead of a teleological destination, then the South Pacific region has put some institutional structures in place that will work favorably in this process. Moreover, the region has connected the living resources of the oceans around it to its own security and well-being.

> The consequence is that policy-makers and the general populace regard the condition of the surrounding oceans and the state of their living resources as probably the most critical issue of national and regional economic importance. Constrained by physical and economic factors, and conscious of their shared marine environment, the Pacific Island States and Territories have utilized regional cooperation, largely through the establishment of a number of regional organizations discussed below, to address many of the coastal and marine resources development issues facing the region. (Tsamenyi 1999, 466)

However, a great deal of work in science and politics needs to come together to reverse such threatening storms on the horizon as apparent increasing poverty, income inequality, political instability, and resource depletion.

I include American Samoa, Australia, Cook Islands, Federated States of Micronesia (FSM), Fiji Islands, French Polynesia, Guam, Kiribati, Marshall

Islands, Nauru, New Caledonia, New Zealand, Niue, Northern Mariana Islands (CNMI), Palau, Papua New Guinea (PNG), Pitcairn Islands, Samoa, Solomon Islands, Tokelau, Tonga, Tuvalu, Vanuatu, and Wallis and Futuna in the South Pacific region (Secretariat of the Pacific Community 2003). France, the United Kingdom, and the United States of America are also members of the Pacific Community and still hold strong influences in the region.

This area falls within the West Central and Southwest Pacific areas with the United Nations Food and Agricultural Organization (FAO). Often, literature on the region separates out the small islands and refers to this group as the South Pacific island nations. Like the other areas studied here, the South Pacific area is incredibly diverse politically, culturally, and historically. There are more than 2000 languages spoken in this region which spans 30 million sq. km, from the Pitcairn Islands in the east and Papua New Guinea in the west (South Pacific Regional Environment Programme 2003). However, there are shared histories and some shared values that span these areas since before European contact (Jackson 1993).

Historical Connections to a Globalization

Indigenous people of the South Pacific have commonly thought of the ocean as precious global and universally interconnected resources which can be used but must be protected in a way to sustain life that depends on it (Jackson 1993).

Occupation in the South Pacific ranges wildly between relatively recent (3,500 years or so for the small islands [Lobban and Schefter 1997]) and ancient (40,000–60,000 years for Australia in particular [Walker 2002]).[2] The colonial period started in the South Pacific in the seventeenth century; during the period before European contact, island peoples of the South Pacific had regional contact with one another. This is a set of cultures that have overarching South Pacific threads shooting through them.

Here I rely on Wartho and Overton's (1999) account of the region's history. In 1565, Spain claimed Guam, but did not send inhabitants until a hundred years later. The beginning of the 1700s saw an influx of imperial annexations, and by 1906, all but Tonga was governed colonially (Wartho and Overton 1999). Thus, the region's interaction with the first wave of globalization was important, but unlike in Southeast Asia and the Caribbean, it came later and was arguably less intensive, probably due to its distance from the rest of the world and insulation by the Pacific. Nonetheless, this period of globalization plays an important role in the current form of economic expansion and trade now being experienced as the third wave of globalization since the end of World War II. In fact, Wartho and Overton argue that the primary determinant of higher per capita income countries in this region and those with lower per capita income is aid from the metropolitan countries to their former or current colonies, aid which is diminishing in amount but increasing in "strings attached."

Several South Pacific islands are dependent territories. American Samoa and Guam are dependent on the United States; French Polynesia, New Caledonia, and Wallis and Futuna Islands are governed by France; the Pitcairn Islands are governed by the United Kingdom; and Tokelau is held by New Zealand. There are also semiautonomous states which still have colonial ties, such as the Cook Islands and Niue, which are tied to New Zealand, and the Federated States of Micronesia, the Marshall and Northern Mariana Islands, and Palau, which are tied to the United States. This makes getting information on these islands very difficult.

> Thus, the colonial era in the Pacific is not over. Although the transition to independent sovereign states with largely democratic systems of governance has dominated the politics of the region for the past 30 years, it is likely that a number of states will choose to remain quasi-colonies, with continuing close ties and support from their metropolitan patrons. (Wartho and Overton 1999, 37)

While trade flourished within the Pacific Islands prior to colonial contact, "it was the colonial era that led to the incorporation of the region into the evolving global economy. Products from the region—copra, gold, coffee, sugar, nickel, timber and phosphate—were exchanged for the plethora of manufactured goods from Europe and North America" (Wartho and Overton 1999, 38). As aid decreases, countries are expected to turn more toward their forest and fishery resources, in addition to tourism, for revenue.

Most Pacific island countries (PIC) are subsistence based. Tables 4.1 and 4.2 and figure 4.1 show the intensity of trade (amount) is low, but that Australia is clearly the trade hub in the region. Much of this trade is narrowly focused on tuna. As in the early period, economic globalization has not reached the Pacific Islands and the larger region, as it has in Southeast Asia, Europe, North America, or even the Latin America and Caribbean (LAC) region (chapter 5). Data on this area is poor. Most countries in the region are not "counted" in the World Bank data on FDI and other measures, which I take to mean that these uncounted countries are not of vital interest, or they do not report this information, to the World Bank. However, these are not countries known for secrecy, and detailed accounts of these economies seem to indicate that this data is representative. Thus, it seems safe to assume that capital intensity is low in general in the region.

Table 4.1. Level of Globalization in the South Pacific

Country	Foreign Direct Investment 1988–1990 (inflows, millions of $US)	Foreign Direct Investment 1998–2000 (inflows, millions of $US)
Region	1,899 (average)	1,981 (average)
Region without Australia	479 (average)	537 (average)
Australia	7,582	7,758
Fiji Islands	44	25
New Zealand	1,693	1,937
Papua New Guinea (PNG)	171	179
Solomon Islands	8	10

Source: Adapted from World Resource Institute 2003a/b.

With lower capital intensity, the South Pacific is more comparable to the LAC, but these areas have important differences. The LAC region and the South Pacific now experience about the same amount of foreign direct investment, if Australia is included and the United States excluded. Foreign direct investment is the flow of money to acquire a management interest of 10 percent or more of a company in another country, and is a frequent measure of economic global expansion as well as national economic health.

However, once Australia—the region's hub of economic activity and globalization—is taken out, the degree of FDI drops from about $US2 billion to about $US500 million in 1998–2000, which is one-third of the Caribbean's FDI with-

Table 4.2. Corporations, Debt, and Assistance in the South Pacific

Country	Parent Corporations	Foreign Affiliate Corporations	External Debt (% of GNI)	Official Development Assistance Receipts (% of GNI)
	1994–2000		1998–2000	
Region	—	3,172	—	—
Region w/out Australia	—	633	43 (average)	8.8 (assumed to be higher)
Australia	610	2,539	—	—
Fiji Islands	—	151	10	2.2
New Zealand	—	81	—	—
Papua New Guinea (PNG)	—	345	69	7.9
Solomon Islands	—	56	50	16.5

Source: Adapted from World Resource Institute 2003a/b.

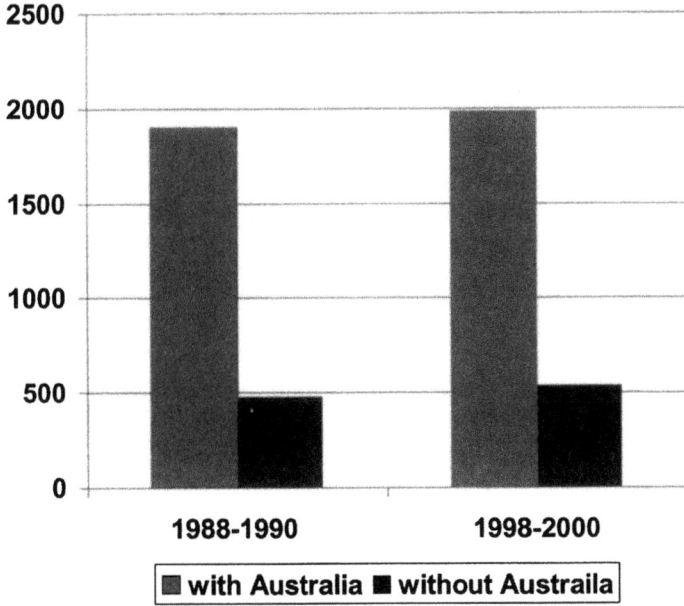

Figure 4.1. Change in Foreign Direct Investment (FDI) in the South Pacific, 1988–2000, with and without Australia
Source: Adapted from World Resource Institute 2003a.

out the United States. With or without Australia, FDI in the region remained nearly unchanged throughout the boom of the 1990s, which supports the conclusion that the reach of capital and markets in the Pacific Islands has not been as intensive as it has been in the rest of the world.

Ecology

Fisheries

In 2000, the South Pacific region produced 15.8 mmt of fish and fish products. FAO data for the Southwest Pacific statistical region 81 show it is one of two areas in the world where there are no overexploited, depleted, or recovering stocks. In other words, all of the fish stocks are estimated to be underexploited (about 4 percent of worldwide fish stocks), moderately exploited, or fully exploited. The other region in this situation is the western Indian Ocean region on the eastern coast of Africa, which neighbors the eastern Indian Ocean statistical area.

Table 4.3. South Pacific Fisheries

FAO Designation	Pressure	Trajectory of Catch	Biodiversity	Other Signs about the Fisheries
About 8% overfished; 92% under-, moderately, or fully exploited in western central region; none overfished in southwestern region	Catch doubled every 5 years since mid-1970s until 1991	Leveling off and slightly declining since 1991	Indications of serious fish-related biodiversity are not clear at this point, reef fish and sharks show serious decline	A few cases of overfishing such as in the Patagonian toothfish and orange roughy

Source: FAO 2002; FAO 2003a; FAO 2003b; Clark 1999.

On the other hand, the western Central Pacific region, where the vast majority of Pacific Island nations are located, is in a similar position but does have about 8 percent of its fishing region overexploited, depleted, or recovering (FAO 2003a).

"World marine capture fisheries production increases in 1999 and 2000 came mainly from fisheries in the Southeast Pacific. Landings from these fisheries grew by 77 percent in 2000, following a marked decrease of 44 percent between 1997 and 1998" (FAO 2002, 8). Most of this increase came from a record-level fishing effort of tuna species.

The region provides about 50–70 percent of the world's tuna for canning and about 30–40 percent of Japan's tuna for sashimi, averaging a total of about 2 mmt/yr. Skipjack and yellowfin tuna species are moderately exploited; bigeye are fully exploited, but not overexploited unlike some temperate tuna species in the Atlantic, probably as a result of regional cooperative binding agreements to limit fishing efforts to 1993 levels (Reid et al. 2003).

Recruitment of tuna is tied to climate variation. The principal fisheries scientist for the Secretariat of the Pacific Community notes that the warming waters provide alternating effects of recruitment depending on tuna species: "The extension of the warm waters in the central Pacific during El Niño events that extends the skipjack spawning grounds may conversely reduce those of the albacore" (Lehody, Chai, and Hampton 2003). This effect has been confirmed in 2004 where colder waters of La Niña (1999–2001) provided more and larger South Pacific albacore at four to five years of age (Lehody, Personal Communication 2004).

Other species are being overfished in the area; for example, orange roughy has seen a rapid decline (Clark 1999). A 1997 FAO report noted that before the end of the 1960s, almost all fishing was done for domestic consumption as an alternative to more expensive red meat. After the 1970s, the global market began focusing more on exporting fish from the region.

> The main species caught during the 1950s and 1960s were the New Zealand dredge oysters, golden snapper, tarakihi and red rock lobster, all caught by New Zealand. These four species made up over 40% of total landings until 1970. . . . By the mid 1970s catches of species from what are now considered the main ISSCAAP (International Standard Statistical Classification of Aquatic Animals and Plants) Groups had already trebled, and catches in total more than doubled. This rate of increase continued, with catches of the main ISSCAAP Groups nearly doubling every five years until 1991, at which point declines in catches of several of the Groups occurred. (Shotton 1997, online)

Thus, the South Pacific fisheries are somewhat mixed. The FAO's assessment that it is the one region that has not overfished any specific fisheries is balanced out by the above account that declines have been occurring to some degree since 1991, and the feelings by locals that they have to spend more effort to catch the same amount of or fewer fish (Young 2004).

Coral Reefs

About 25 percent of the world's reefs exist just within the western Central Pacific FAO statistical region (Sea Around Us Project, 2004). Reefs in the South Pacific are among the most biodiverse in the world: "Reefs in the central Indo-Pacific have more than 10 times as many coral and fish species as reefs of the Galapagos Islands in the eastern Pacific" (Knowlton 2001, 1493). The majority of coral reefs in the South Pacific region are doing well or face a low likelihood of mortality in the next few decades. Reefs which are not doing well are threatened by rising temperatures and nearby urban development. Also, reef fishing and the crown-of-thorns starfish, which eats coral, threaten these area reef systems.

Table 4.4. South Pacific Coral Reefs

Level of Risk	Low	Medium	High
Proportion of Reefs at Risk (%)	59	31	10
Global Reefs at Risk (%)	42	31	27

Source: Spalding, Ravilious, and Green 2001.

Sea Level Rise and Marine Temperature Change

Sea level rise is one of the preeminent threats to the South Pacific given the area's vulnerable small island states which are now watching the shoreline come closer and closer; however, the South Pacific is fortunately seeing less sea level rise than the global mean of 2 mm/year. According to Wong and others (2001),

> Overall, the area-weighted steric height change for the North and South Pacific oceans from 60°N to 31.5°S, based on the four zonal sections, is estimated to be a rise of 0.85 mm yr^{-1}. That is, because of significant ocean warming from the subtropical North Pacific, the overall steric sea level in the Pacific has risen in the roughly 20-yr period between the 1960s and 1985–94. (1630)

"Steric" refers to the thermal expansion of the water. Thus, this answers two of our questions about South Pacific oceans and climate change. First, the water is warming slightly at the higher portions of the water column, but is remaining relatively stable in the deeper section. However, the water column itself is changing in salinity and the depth at which the lower-salinity/higher-temperature water flows. Second, the sea level is moderately raising, on the whole, less than half as much as the other two regions or the global average of 2 mm/yr. This assessment agrees with data gathered by satellite altimetry (Nerem and Mitchum 2000). Some countries in the region are in more trouble than others depending on their geological character. Parts of Fiji—the Vanuabalavu Island group, for example—is subsiding, that is, the landmass is sinking at the same time the sea level is rising (Nunn et al. 2002). Vanuatu, on the other hand, is rising at about 3 mm/yr (Cabioch et al. 2003).

Water temperature is both warming and cooling in the area. Ice melt is "freshening" the subpolar water in the region, and this process cools that water. In turn,

Table 4.5. South Pacific Climate-Related Ocean Changes

	Experienced since 1960s	Projected
Sea Surface Temperature Change from the Mean	+0.12°C on average, with some areas actually cooling; the warmest section at 24°N with +0.1701°C	+0.4°C extrapolated from .008°C/yr
Sea Level Rise	.85 mm/yr	+5 cm next 50 years

Source: Wong 2001.

this changes the salinity-temperature mix which actually results in warmer water in the South Pacific region. Thus, "a trend in deep warming is apparent in repeat hydrographic sections across the subtropics from 1957 to 1981 to 1992" (Johnson and Orsi 1997, 306).

Also, under climate change, as modeled by El Niño Southern Oscillation events (ENSO), tuna may be displaced by as much as 50° latitude (Hunt 2003), which will place much of these tuna outside Exclusive Economic Zone (EEZ) regulation, in addition to the species-specific changes noted above for skipjack and albacore. If this does indeed occur, then small island nations in the Pacific will experience a dramatic loss in relative income, and the fish populations will lose some of their regulatory protections, particularly if they do not cross any EEZs. Further, tropical cyclones are expected to be more severe (not necessarily more numerous), and droughts are expected to frequent the area (Hunt 2003).

Reefs will provide to the Pacific region some protections from the storms, as well as some sea level rise. This means that the region needs to have in place protections for the reefs, which in turn means governing fishing and sediment deposits which result in damaged reefs—all in order to adapt to climate changes. At the same time, institutional ability to protect these components of the regional coasts from maladaptive fishing practices (e.g., cyanide and dynamite usage) and coastal development which increases sediment and pollution is extremely vulnerable and weak.

Regional Political-Economic Conditions

Poverty and Violence

Data about the region's poverty is inadequate; the World Bank has no available poverty data in its development indicators. Even the larger states in the region have unknown (or unpublished) degrees of poverty. This gap forces our estimation of this factor from a source other than the World Bank. The data provided in table 4.6 are from the Asian Development Bank, which determined the poverty line "based on basic needs." Due to the ambiguity of this designation (UNESCAP 2003), the available poverty data is not strictly comparable to the World Bank poverty data used in the other two regions discussed in this book. Nonetheless, it is clear that poverty in the island regions is serious and becoming worse. ADB and the United Nations Economic and Social Survey (UNESCAP) attribute this situation to a reduction of economic growth. The report also notes that this is a problem for ecological sustainability because it is obvious that much of the economic growth seen in Asia has come with direct ecological costs. The region's apparent high degree of poverty is, based on the average, ten points higher than in Southeast Asia and fifteen points higher than in the Caribbean, and indicates one of the most serious threats to the region's sustainability. The report describes that Papua New Guinea had rising population, poverty, low soil fertility, low access to education and health care, and a rise in periurban informal settlements that were unstable and unsafe (UNESCAP 2003).

The difficult question here is obvious: is this poverty a result of increasing liberal pressures or of the fact that the liberal policies have not been fully embraced? The response is probably not so clear cut between these two options; however, as this section describes, if the modern nation-states are to be retained, they require revenue-based funding, not subsistence-based bartering and trade.

Other indicators are more promising. We can see from the data set out in table 4.6 that military expenditures are far outpaced by expenditures for social goods like health and education. I take this as a good sign, and I would like to interpret this as a reflection of the preference for nonviolence found in some of the traditional cultures of the region and a spilling over of this preference into the formal governments of the South Pacific, perhaps symbolized by the regional nuclear weapon ban policy. This may or may not be the case, but we can see that average government expenditure on these social goods is relatively high compared to that than in any other region in relation to military spending. The information is not strong enough for any serious claims about this comparison, but it is at least in keeping with the fact that this region has experienced fewer neoliberal pressures that would tend to decrease the social expenditures, and it is in keeping

Table 4.6. Poverty and Government Expenditure in the South Pacific

Country	% of Population Living in "Basic Needs Poverty" (ADB)	% of Population Living on Less than $2/day	Military Expenditure (% GDP) 2000	Health Care Expenditure (% GDP)	Education Expenditure (% GDP)
Region (average)	29.42	—	1.2	4.4	6
Australia	—	—	1.7	6.0	4.8
Fiji Islands	25.5	—	—	2.9	—
New Zealand	—	—	1.0	6.3	7.2
Papua New Guinea (PNG)	37.5	—	.8	2.5	—
Samoa	20.3				
Solomon Islands	—	—	—	—	—
Tonga	23.8				
Vanuatu	40				

Source: Adapted from World Resource Institute 2003a/b.

with my findings that the region is more sustainable than others, assuming these social expenditures can be taken as rough proxies for commitments to comprehensive social well-being and security.

Violence in the region, measured at the armed conflict level, was zero from 1989 to 2000 (Eriksson, Wallensteen, and Sollenberg 2003). There are a number of conflicts not counted in this data, though: border tensions between PNG and Indonesia in 1984; the secessionist movement in Vanuatu in 1980; the independence-related violence in New Caledonia in 1984–1985 and 1988; coups in Fiji in 1987 and 2000; Bougainville's secession attempts from PNG starting in 1989; and ethnic violence in the Solomon Islands (Rolfe 2001). The Bougainville revolt had at least 12,000 casualties over nine years. The revolt was tied directly to globalization first when Bougainville was included within PNG instead of the Solomon Islands during colonization, and when the PNG allowed an Australian mining company to bring 10,000 workers from around the world into an indigenous community of 14,000 people. The mine devastated traditional social arrangements, and the "Pollution was so extreme that river systems suffered irreversible damage. Consequently, local agriculture was devastated" (Rolfe 2001, 42).

That said, this region is unlike the other two in that there were no conflicts to qualify in the Eriksson data. On this important level, violence meets the criteria for Borgese's nonviolence, though there is obvious and important work that remains to create a regional nonviolent set of societies.

Fishing

The fishing industry provides Pacific islanders US$262 million and about 39 percent of their protein (World Bank 2000). The vast majority of Pacific islanders also feel that the catch per unit of effort is declining (World Bank 2000; Young 2004). To the degree that these small fishers use their efforts for subsistence and minor trading, the increasing difficulty in catching fish also increases their food insecurity. This is a grim trend coupled with the increasing poverty felt in many places all over the world as prices rise and the ability of the poor to buy fish becomes less tenable (FAO 2002).

Oceania holds only 0.2 percent of the world's decked fishing vessels, as opposed to the neighboring region of Asia, which has almost 85 percent of these typically commercial fishing boats (FAO 2002). The vast majority of fishing pressure comes from outside the region by foreign fleets, at the same time that the region has minimal patrolling and enforcement capabilities. Most of the tuna is skipjack caught through purse seine ships operated by the Federated States of

Micronesia, Japan, Korea, Papua New Guinea, the Philippines, Spain, Taiwan, the United States, and Vanuatu; nearly all of the tuna is shipped to Thailand for final export to affluent countries (Reid et al. 2003). As of 2000, two-thirds of these nations and 65 percent of the Gross Registered Tons in purse seine fishing capacity are from outside the region (Reid et al. 2003).

The issues of fisheries are of vital economic interest in the Pacific. In order to protect the region's fisheries, especially its tuna stocks, the sixteen area countries in the South Pacific Forum created the Forum Fisheries Agency (FFA) in 1979, headquartered in the Solomon Islands. New Zealand and Australia are represented here, but not France or the United States. The South Pacific Forum is primarily an agency for political and economic cooperation, and thus the FFA retains those concerns. For example, the FFA keeps and provides transparent information on the tuna markets; however, tuna sustainability is not as highly visible as a concern on their website. In contrast, the prominence of sustainability is evident in Pacific Community Internet documents.

Exports of tuna generated from migratory stocks are responsible for more revenue than any other source in Pacific economies (Hunt 2003). Before 1980, the tuna industry was nationally integrated so that tuna caught by U.S. fishers was consumed in the United States (Bonanno and Constance 1996). After that time, prices for tuna declined and the industry restructured through transnational arrangements now commonplace in the global economy. Even so, in 1987 ten major corporations dominated about 70 percent of all tuna and tuna products including the areas of catch, shipment, processing, distribution, and trade (Bonanno and Constance 1996). Bonanno and Constance list these ten corporations as Bumble Bee, Mitsubishi, Marubeni, Mitsui, Safcol, StarKist, H. J. Heinz (Australia), Van Camp (Chicken of the Sea), C. Itoh, and John West. At that time, the U.S. tuna market leader was StarKist (H. J. Heinz of the United States), which held about 30 percent of the U.S. tuna market, and included the world's largest and second-largest processing plants in Puerto Rico and American Samoa (Bonanno and Constance 1996). Van Camp Seafood (Chicken of the Sea brand tuna) had the second-largest share of the U.S. market with about 16 percent (Bonanno and Constance 1996).

Now, Del Monte owns StarKist, which controls 45 percent of the *world* tuna market. Bumble Bee controls 22 percent, and Chicken of the Sea (owned by Thai Union Frozen Products as of December 2000) controls 17.5 percent (Chanjindamanee 2000). Thus, globalization has fragmented the structure of how tuna is caught, processed, shipped, and sold, but the tuna market is more concentrated with fewer companies controlling more of the market. In terms of gross production, Thai Union Frozen Products became the second-largest tuna company in the world (Chanjindamanee 2000), which may account for Thailand's being the largest exporter of fish products in the world in 2001. This concentration has some correlation to increased fishing pressure in the region.

The tuna fisheries (bigeye, skipjack, yellowfin, and albacore) have seen record catches since the late 1990s. The years 1998 and 2002 saw the highest annual catches on record for the region, providing for 50 percent of the world's tuna in 2002 (Williams 2003).

The fact that the fisheries in this area depend on development level affects projections twenty-eight years into the future. Developed countries expect a 5 percent increase in capture production and a 1 percent increase in aquaculture production. Developing countries expect a zero percent increase in capture production and zero increase in aquaculture production through 2030 (FAO 2002).

However, licensing out the zones where the tuna is caught will likely remain very important. Receipts from licenses for access to their large EEZs make up the lion's share of government funds for some small island nations; Tuvalu, for example, receives about 50 percent of its government revenue from foreign tuna-fishing fleets (Hunt 2003).

The South Pacific Project Facility, a subgroup of the World Bank, is responsible for capitalizing South Pacific fleets to exploit their own resources (Hunt 2003). This is a double-edged sword. Development of the regional fishing fleets means that regional fleets are competing against foreign fleets in their own EEZs and on the high seas where no license fees are collected, and the fish populations may be suffering from this double exploitation. Indeed, until 2002, the combined South Pacific fleet did not rank among the top purse seiners for its own region. The purse seine (a type of net employed by large industrial fishing vessels) catches about 55 to 60 percent of the Western Pacific Ocean fish per year.

Taiwan, Korea, and Japan have the largest purse seine catches in the region (Williams 2003). This is an improvement over the incredibly marginalized position of Pacific countries, but it still indicates that the foreign fishing industry wields a very heavy hand in the fisheries and the region. What is worse is that in order to improve the South Pacific position, these countries feel they have to industrially participate in the fisheries in the same fashion that foreign fleets do, thereby distorting and degrading the small-scale and historically sustainable fishery management of traditional practices.

This process, according to Salome Samou (1999), began in the late 1950s when the first cannery was established in American Samoa. Clearly, now that it has its own fleet, the Pacific Community is participating more in its own fisheries, but, Samou writes, "The distant-water fishing nations dominate the industry whilst Pacific states have little power or ability to develop their own industries"

(148). Also, these nations have historically caught only 6.5 percent of the tuna in weight in the South Pacific Community statistical area (Adams and Majkowski 1997). Indeed, the United States did not even pay access fees to these countries until they threatened to align with the Soviet Union in the late 1980s (Schurman 1998). The most important distant-fishing nation at that same time, Japan, refused to pay more than 3 percent of the estimated value of the catch (Schurman 1998), which apparently set the stage for current payments. Thus, in producing half of the world's canned tuna and 30–40 percent of Japan's sashimi, by 1995 "the Pacific Islands received only $US60 million in license fees, the equivalent of 3.53 percent of the catch" (Samou 1999).

In addition, distant-water fishing is heavily subsidized to the extent that capitalization of fishing fleets in the European Union is bolstered by $1 billion, and the U.S. fishing fleets by $14 million; these fishing fleets are finding their way to the Pacific to catch extremely valuable tuna. For example, Spanish subsidization added 14 "super seiners" to Kiribati's EEZ (Hunt 2003). Nonetheless, much of the institutional development has occurred as a result of affluent countries' contributing to regional organizations, such as the Pacific Community and the Forum Fisheries Agency, where most of the fishing governance is occurring.

Pressures to the tuna are only one area of concern. There is no regional management of shark catch, which was much higher than tuna catch (400 tons in 1999), and turtle catches are also of great concern. However, on the regional level, these issues receive far less attention than the economically important tuna, even though serious ecological harm may be occurring.

Coastal Development Projects

Coastal development projects in the South Pacific are notably degrading the marine environment, and have done so since European settlement in the region. For example, European land management practices of timber clearing and overgrazing increased the sediment load in rivers to the Great Barrier Reef between five- and tenfold, whereas prior to the settlement era, heavy sediment rarely made it to the reef—this alteration combined with anthropogenic forced climate changes have had a "severe impact on river catchments and hence on the nearshore coral reefs of the GBR" (McCulloch et al. 2003, 730). This damage continues today through extended coastal developments:

> Terrestrial development may however be an even more serious threat to Pacific inshore resources. Export logging in the Solomons and Papua New Guinea is in theory subject to codes of practice, but uncontrolled logging and logging on steep slopes causes soil erosion that affects inshore waters

and coral reefs. Likewise the run-off from export agriculture such as squash in Tonga and oil palm in Papua New Guinea, can pose threats to inshore resources. The potentially deleterious effects on marine resources of the disposal of export gold mining tailings to the ocean bed in Papua New Guinea is causing disquiet, despite the claims of mining companies. In each of these cases the export activity is a function of foreign capital inflows embodying technology. (Hunt 2003, 82)

In the Solomon Islands, Malaysian logging companies, such as the notorious Marving Brothers Company, were found to have bribed officials with several million dollars cash and other benefits; at the same time, logging in the Solomon Islands was well beyond the sustainable cutting level. Marving Brothers and the Chinese company Kyuken Timbers have both been separately implicated in the killings of Martin Apa and Sony Tong, both local lumber operations opponents (Scheyvens and Cassells 1999; Roughan 1997). These logging operations on the Solomon Islands have been directly responsible for some deaths of coral reef.

Vanuatu had placed a moratorium on logging, but soon revoked the law, giving exclusive logging rights to five companies (Roughan 1997). This change implied a strong concentration of power in these companies within the government. In Papua New Guinea, the government has failed to properly enforce timber regulations, "which has worked to subvert the interests of the indigenous landowners" (Roughan 1997, 112).

Of course, not all of the foreign money coming into the South Pacific is having this effect. There are some World Bank and Asian Development Bank collaborative projects aimed at improving marine area management on the coast. The Marine Biodiversity Protection and Management Project in Samoa is one such project, funded mostly by the World Bank. Other contributors include AusAID, the government of Samoa, the local impacted community, and the World Conservation Union (IUCN), which is implementing the project and hopes to halt the loss of both coral reefs and biodiversity in the area through local inclusion of marine and coastal management (World Bank 1999).

It appears that systematic data on coastal development, as in all three of these regions, is lacking; it also appears that the anecdotal evidence indicates that coastal-development projects in this region have not had the same level of impact and toll as they have in the Caribbean or Southeast Asia.

Knowledge Production

Indigenous knowledge systems about South Pacific ecology are known to have been prodigious. Ecologist Robert Johannes has studied the knowledge of the "old coastal culture" in the Pacific Islands, noting that the islanders "seemed to know a lot about marine biology that we didn't" (quoted in Safina 1998, 311). These fishers are becoming more rare as their children pursue different paths and the transmission of this knowledge is lost in light of the pressures of industrialized fishing and the movement toward wage labor (Safina 1998). These are very real pressures against the people who hold this information, which is likely built on hundreds of years of experience.

According to scholar Greg Fry (1997), knowledge production in the South Pacific has been framed by and reflects the interests of Australia, which he believes has seen itself as the leader of the postcolonial region through a form of "Australasian Monroe Doctrine" and manifest destiny. Australia approaches natural resources in the same way that other "settler capitalist" countries, like New Zealand, do: it extracts natural resources for transfer out of the region. This method has come with dire expense to indigenous peoples and the ecology of this country (Walker 2002). This may explain why Australia is the hub of capital in the region, and is one of the only industrialized countries in the area besides New Zealand (see table 4.1).

Evidence form Fry and other scholars such as Epeli Hau'ofa (1994) indicates that the knowledge led by and produced under Australian fiat has actually provided the "basis" for the regional agenda. Thus, the ability of the center of capital for this region to control how the region is viewed has been notably powerful. However, this ability is not all encompassing.

Regional Institutions and Civil Society

"Almost every basic fisheries conservation measure devised in the West was in use in the tropical Pacific centuries ago" (Johannes 1978, 352). It is likely that this history and knowledge of ocean management and ecology has transferred in some way to the modern Pacific regional institutions.

The South Pacific has developed a number of institutions organized around the region's ocean resources. Institutionally, centralized Pacific enforcement of limits to fishing, particularly inshore live fishing, is weak since these agencies are usually underfunded. For example, it would take hundreds of years simply to do simple surveys of reef fish abundance, and island countries cannot financially or environmentally afford such time and effort. This is why many areas are moving to "dataless" management, which relies on rules of customary marine tenure and which has worked as well as—if not better than—modern science-based approaches (Gillett 2002).

There is a strong historic precedent for comanagement of ocean resources, which included communal responsibilities and was sustainable. There is increasing recognition by the island governments that these traditional institutions, such as the rule of taboo that installs a no-take marine area, need to be emphasized. Island nations in the southwestern Pacific all note declining catch in fisheries and mollusk, and are considering returning ownership of coastal areas to the village level in order to reinstate and reemphasize the ability of these practices to protect marine resources as "tacit acknowledgement that western-style centralised fishing regulations are failing to protect many of the world's marine ecosystems" (Young 2004, 9). It appears that modern institutions can work off this legacy of traditional practice if it decentralizes coastal management (Doulman 1993). I believe this legacy is one reason that the region is more sustainable than the others which have not had the same protection from globalization. Unfortunately, most traditional practices have been and continue to be eroded by modern institutions, which "have generally failed to prevent overexploitation" (Doulman 1993, 108).

I will now describe the formal regional institutions, using the work of Martin Tsamenyi (1999) and primary information provided by the institutions themselves. Five organizations are responsible for the management of these resources: the Pacific Community, the Forum Secretariat, the South Pacific Forum Fisheries Agency, the South Pacific Applied Geosciences Commission, and the University of the South Pacific (Tsamenyi 1999). Here I discuss the first three in addition to the South Pacific Regional Environmental Programme (SPREP).

The Pacific Community, formerly the South Pacific Commission, is the oldest regional organization in the area, originating in the 1947 Canberra Agreement coordinated by colonial powers. The Secretariat of the Pacific Community provides open access to a great deal of regional issues; is a source of primary knowledge and data of ecosystems—particularly tuna and billfish; and organizes policy and coordination on a wide array of topics, including development, public health issues, socioeconomic conditions, and women's issues (Secretariat of the Pacific Community 2003). This regional organization is one of the more important formal organizations in the region (comparable to CARICOM and ASEAN in the other two regions), and an analysis of the content of its primary missions indicates that sustainability is among its top rhetorical priorities. It notes on its website (www.spc.org) that its mission is to "develop the technical, professional, scientific, research, planning and management capability of the Pacific Islands people to enable them to make informed decisions about their future development and well-being." The treatment of the organization's programs of land management, marine resources, and socioeconomic resources indicates a diversity of socioecological interests, where economic well-being is one among other issues that create "well-being." I take this to mean that the neoliberal programs of the current episode of globalization do not create the functional base of this region. Export and economic growth are not unvalued, but they are not dominant either.

The South Pacific Forum is an organization which excludes France, the United Kingdom, and the United States, specifically to favor the interests of the region. Being more independent than the Pacific Community organization probably explains why the South Pacific Nuclear Free Zone Treaty (also known as the Treaty of Rarotonga) originated with and was signed by the parties of this group (Tsamenyi 1999). Following similar requirements set out in the 1967 Treaty of Tlatelolco (Suter 1996), this treaty bans stationing, testing, using, or threatening to use nuclear weapons in the zone. The forum continues to call upon the United States to ratify this treaty which it signed in 1996, but this is unlikely since the treaty was negotiated and signed in 1985 and the United States does not favor "limiting its nuclear options" (James Lilley quoted in Suter 1996). This group, though not without some members dissenting, also supports the continued protection of whales through whale sanctuaries (Pacific Islands Forum 2003). Finally, this institution advocates for greenhouse emissions limits within the Framework Convention on Climate Change both within its region and globally, albeit with little if any success outside its own members. Consequently, the group also collaborates on regionwide and country-level adaptations that will be necessary in the face of rising sea levels. The institution sees outside pressures on the climate as mostly beyond its control.

The Forum Fisheries Agency (FFA) is concerned with harmonizing fishing requirements at the country level and aiding in cooperation between distant-water fishing nations. It has been successful in putting together a regional register of foreign fishing vessels, harmonizing minimum requirements for foreign fishers, and establishing a regional treaty with the United States, as well as a regional ban on driftnets (Tsamenyi 1999). It is thanks to the initiative of the FFA that the South Pacific is one of the first regions to organize a treaty under the United Nations Fishery Stock Agreement (FSA). On June 19, 2004, the Convention on Conservation and Management of Highly Migratory Fish Stocks in the Western and Central Pacific Ocean went into force. This treaty regulates migratory and straddling stocks at the regional level with the ability to coerce and enforce foreign vessels not adhering to the fishery limits of the region (Western and Central Pacific Fisheries Convention Preparatory Conference, 2004). While this treaty provides new opportunities and a strong institutional mechanism to end the important CPR fishery problem in the area, monitoring of the zone will take development assistance and even more regional cooperation (Ram-Bidesi and Tsamenyi 2004).

SPREP's principal charge is to coordinate sustainable management of coastal and marine resources in the region, but one of the things that makes SPREP a jewel in the struggle for global sustainability and South Pacific efforts in particular is its knowledge-production programs. SPREP not only develops knowledge about social-ecological connections, it has programs that disseminate specific local knowledge and train local individuals, organizations, and government officials to do *their own* environmental assessments. The organization is heavily influenced from abroad, but the programs in place are meant to build local capacity and protect local social-ecological conditions. The United States funds 21.5 percent of SPREP's general budget, France pays 15 percent, and Australia pays 22 percent, leaving the remaining 41.5 percent for the smaller countries (SPREP 2002, 80). Therefore, this transregional funding is one example of global finance that has enriched the lives of locals and runs contrary to my general findings of the impacts of economic globalization. Perhaps this is because even though neoliberalism is encroaching upon the area, economism and FDI influence remain balanced against other goals.

SPREP also uses Action Plans, and is an implementation agent for several international environmental laws, such as the Basel Convention (limits to the transportation of hazardous waste) and the London Convention (limits to dumping hazards at sea). SPREP utilizes both national and subnational levels to improve environmental governance and knowledge, with the specific goal of Pacific independence. For example, in its "Pacific Futures" program, it is working with island communities to deal directly with climate change and its accompanying sea level change. The program includes the generation of the residents' own knowledge about regional climate change–related problems.

Regional politics, commerce, and culture of the South Pacific and the sustainable practices with which they struggle are increasingly subject to Western influences, a situation that is resulting in deeper gender inequity. These inequalities are explained by a switch from subsistence to wage-based income.

> As the economies of the Pacific Island countries have become more monetised and men have taken on more paid work than women, women's contributions to development have not been accorded the respect they once received. In some cases, women are no longer consulted on important issues of resource extraction or development which will impact directly on their livelihoods. (Scheyvens 1999, 53)

Further, Scheyvens and Lagisa (1998) have found that this deepening inequality is behind acts of resistance from local sabotage to civil war. Thus, institutions and economies of the Pacific are marginalizing women in a way that threatens sustainability, and this problem has worsened corresponding to the time of increased contact with Western economic demands. Formally, the Pacific Community says that since the early 1980s it has been dealing with gender equality through its Pacific Women's Bureau. The community has at least recognized that rigid gender roles and marginalization are the status quo (Pacific Women's Bureau 2003). Thus, it appears that women are seriously and significantly marginalized in the South Pacific, and that there is a regional institution in the area to address this problem.

However, there are local cases in the region that point to a promising cultural asset of nonviolence. Maria Lepowsky's (1994) research on the island of Vanatinai, now located off New Guinea, shows that there have been serious episodes of violence in the island's history, but there continues to be a "distinct absence of a 'culture of violence'" (Lepowski, 199). In particular, gender-related violence is abhorred and virtually nonexistent. This statement cannot be generalized to the region (Birkett 2004), but I believe that the fact that such an example exists anywhere is important, and I would imagine it informs the regional politics in some oblique way.

Comparison to the Borgese Test of Sustainability

Traditional indigenous law and practices and institutions created significant limits on natural-resource exploitation that are directly compatible with the Borgese Test. To the degree that the region is able to use these limits as a guide to survive the increasing liberalization of its political economy (and any potential increases in economic globalization), it may be able to delay some of the declines occurring elsewhere. However, some of these declines, like global-warming related changes, will not be solved regionally, and the changes to the ocean by other regions in this manner show a direct universal connection, for better or worse.

Current institutions in the region give the appearance of serious attention to a broad array of socioeconomic concerns, and show that a "return" to the traditional is not necessary for sustainability, but that the values and rules of traditional South Pacific societies, selected for their impact on overall well-being, are crucial for the region's survival and perhaps for the World Ocean in general. Empowering indigenous communities to lead in this way may produce desirable political and ecological effects. But to the degree that international agreements are tailored with economic utilitarian values embedded in them (let alone neoliberal values), traditional values and laws that place the World Ocean within the "veil of Mother Earth"—which provides physical and spiritual sustenance (Jackson 1993) to the planet—will be structurally weakened and marginalized, even if there are outbreaks and exceptions. According to indigenous law, using the World Ocean as a simple bank of resources to be cashed in is inherently unsustainable because it will not recognize imbalances in the larger system. It is hard to disagree with this position. Thus, the most important policy suggestions that come from this application of the Borgese Test are to emphasize, learn from, and take the lead from indigenous law and institutions that are a wealth of time-honed information about sustainability the region enjoys; and to reverse the marginalization of indigenous peoples in these island societies.

Synthesizing Ecological and Political Structures to Their Global Structure

All of the region's ecological conditions, from fisheries and coral to the conditions of the water (currents and relative sea level) are declining. Fisheries' changing in populations, coral's dying off, and climate-related changes are making their mark in changing current conditions and relative sea level. Thus, looking at the region and comparing it to itself historically indicates that the region's ecological processes and changes are not sustainable.

Compared to its peer groups, all of South Pacific's ecological conditions show less change or decline than in any of the other regions, indicating that these ocean changes are indeed global in scope and extension, but not intensity. Economic liberal globalization of the number of firms, the level of foreign direct investment, or the level of change in foreign direct investment is the lowest seen in all three regions. There also seems to be less general sociological hierarchy, and there are markedly fewer armed conflicts in this region than in either of the other two. Therefore, the concomitant lower levels of globalization, ecological change, and violence in this region indicate that the intensity of historical and current globalization is indeed an important factor in ecological change.

Where the regional ecology is stressed, it appears to be from modern Western and Asian influences which seem to be involved mostly with the South Pacific for tuna and other rich fisheries of the area. These fisheries are not yet overfished, but they do show signs of geometric increases in pressure and capitalization, which come almost entirely from abroad, and go through an increasingly concentrated set of firms that sell the fish on the world canned-tuna and sashimi markets.

Also, while some coastal development projects and urbanization are evidently local phenomena threatening coral reefs, warming waters are a result of climate change. Climate-related changes are one of the clearest instances of foreign impacts. Looking to table 3.4, on the top twenty-five per capita carbon emission countries, and table 3.5, on the bottom total carbon emission countries, we see that the region is nearly absent. Table 3.4 shows that the only regional representative is Australia, and table 3.5 shows Australia, Palau, and Nauru. Thus, to the degree that carbon emissions force oceanic temperature, chemical, biophysical (as in the change in tuna populations and coral), and sea level advances are imposed from abroad—and if Australia is included as important—these changes are most accurately driven from concentration of global capital, seen in the scale of U.S. emissions, and the per capita usage of other industrialized centers of trade, including Australia as a hub of economic activity (table 4.1).

Notes

1. Remember that I use Borgese's definition of sustainability: the evolution of nonviolent governance accountable to multiple levels of human organization ensuring human material equity, productive ecologies through interdisciplinary knowledge.

2. Remember that I employ Held and others' (1999) definition of globalization, which includes the "transcontinental or interregional flows and networks of activity, interaction, and the exercise of power."

Sustainability in the Caribbean Basin   5

> *It is difficult to find any authority that argues that sustainable development is even clearly in view, let alone close at hand [in the Caribbean]. (Barker 2002, 83)*

EXTRACTION OF RESOURCES from and the use of the Caribbean were "pivotal for the rise of Europe to world predominance," but much of the Caribbean is now relatively ignored or paternalistically seen as a "recipient" region of benevolent aid or advice (Sheller 2003, 1; Sued-Badillo 1992). Not surprisingly, the Caribbean region demonstrates several desperately unsustainable trends including poverty, violence, and ecological catastrophe, and a relatively disempowered polity which began with imperial colonialism starting in the fifteenth century. Despite these problems, when compared to the South Pacific and Southeast Asian peer groups, the region is in the middle of the two respective positions for sustainability.

Included in the Caribbean are countries found in the FAO's Western Central Atlantic Fisheries Commission and CARICOM. Thus, I am including countries from the Caribbean and Gulf of Mexico basins. These countries are Antigua and Barbuda, the Bahamas, Barbados, Belize, Dominica, Grenada, Guyana, Haiti, Jamaica, Montserrat, Saint Lucia, St. Kitts and Nevis, St. Vincent and the Grenadines, Suriname, and Trinidad and Tobago. Barbados, Guyana, Jamaica, Suriname, and Trinidad and Tobago are designated by CARICOM as "more developed" while the rest, except the Bahamas, are designated "less developed" (CARICOM 2002–2004). According to the Human Development Index, Antigua and Barbados also are considered among the "world's nations having the highest levels of human development" (Alonso 2002, 33). Puerto Rico is geographically part of the Caribbean but politically governed through the United States, and most information about Puerto Rico as it relates to the Caribbean is absorbed into the U.S. data. Cuba is part of the region but has a special relationship within the basin as a result of being ostracized by the United States, which has maintained a qualified hegemonic control over the area since the beginning of the twentieth century

(Randall and Mount 1998). Nonetheless, about a third of the island people in the Caribbean live in Cuba, and it is the most industrialized country in the island region; about a fifth live in the Dominican Republic and Haiti respectively, leaving roughly only a quarter of the population in the other islands (Alonso 2002). Of the more than 100 million people in the island region, most live in urban settings (Barker 2002), as in Southeast Asia, though the size of the urban areas and their impact have been less in the Caribbean.

Several Caribbean Sea littoral states in South and Central America have had important influence and are part of the "wider Caribbean region," which includes the above island states as well as Mexico, Guatemala, Honduras, Nicaragua, Costa Rica, Panama, Colombia, Venezuela, and even nonbordering El Salvador (Randall and Mount 1998). Sometimes the entire region is described as the Latin American and Caribbean (LAC) area. I include the United States within the region.

Within the basin, the history of international relations, the variance of government types, and other factors make this region extraordinarily complex. For example, decolonization struggles, the Cuban revolution and the entry of the Soviet Union into the region, the Haitian revolution, the legacy and imprint of slavery, tropical agriculture, a violent drug trade, and the contemporary tourist industry all swirl together with the strength of the well-known hurricanes in the Caribbean.

Physical indicators of the area note intensive deterioration; social and economic indicators are still deeply influenced by the history of colonization; and Haiti—the Western Hemisphere's poorest nation—erupted in violence as I wrote this chapter and has since been embroiled in the controversy of whether or not the United States deposed the democratically elected Aristide. Worse, every year Haiti loses 80,000 children to "hunger and related problems" (Hay Brown 2004, A18).

The Caribbean recently formally organized as a region; this is probably one of the more promising conditions for Caribbean sustainability since this may add some degree of capacity for the governments in the region. The question, though, is who will this capacity be designed for? This question remains unanswered, but the historical and current movement favors free markets, large agribusiness like Dole bananas, and tourism.

History of Globalization

The first inhabitants of the Caribbean islands arrived on Hispaniola and Cuba around 4000 BC, probably from the Yucatan peninsula (Wilson 1997). Over time, they developed a complex hierarchical political organization that grew to include thousands of people within single village areas. Indigenous Caribbean people saw the ocean as a primal and creative force. For example, the Taino (Arawak) viewed the birth of the ocean (*bagua*) and its dependent organisms as the first act of creation (Oliver 1997). Being in the middle of the ocean seems to be part of the reason why the Taino saw themselves at the center of creation, a view which came with "the awesome responsibility of making it work for all generations to come" (Oliver 1997, 152). From this José Oliver laments: "The Tainos are no longer amongst us; their genes have been diluted among the new Old World populations. Their culture—as an integrated holistic system, as a mode of interacting with the natural and supernatural surroundings—is for all practical purposes gone as well" (1997).

The region has experienced centuries of interpenetrating commercial and colonial forces, all of which have used the ocean as a global highway to the islands. Exploitation and export of gold from the Antilles supplied necessary capital to the Spanish empire for other expansionist efforts and was essential to Spain's power at the time (Sued-Badillo 1992). Over a forty-year period starting in 1492, about eight million of the Taino and Carib people, for whom the region is named, were eliminated either by disease, genocide, or slavery (Robertson 2003; Wilson 1997). Nonetheless, the indigenous people of the Caribbean remain a potent symbol of anticolonialism (Joseph 1997).

This first wave of globalization forced habitat destruction and extinction. Deforestation occurred because of the construction of new sugar and banana plantations. Moreover, prior to the first inhabitation, the extinction rate went from one species every 299 years to one species every 122 years. Spanish colonization worsened the extinction rate. In the next 150 years the West Indies, for example, lost a total of 50 species, and many, like the hutia, went extinct within a few decades of contact (Cunningham 1997).

Randall and Mount (1998) document that Spain, Britain, France, and Denmark were the foremost powers in the region from the seventeenth to the nineteenth centuries. At this point the United States became the hegemonic, or overwhelmingly dominant, force in the Caribbean. Early on, particularly through the eighteenth century, the Caribbean "had been the main recipient of the traffic in human cargo from Africa, primarily to work in the sugar cane fields"; the number of workers added up to more than four million Africans stolen from their homelands and enslaved (Randall and Mount 1998, 20).

The trade in slaves was dominated by the English and French. The slaves were shipped mostly through the Dutch, and during this period 90 percent of the African slaves died through forced marches, the middle passage, and field work

(Radford Ruether 1994). After sugar was no longer profitable for the European colonists, the British began to enforce a ban on international trade in human beings in the nineteenth century (Randall and Mount 1998; Radford Ruether 1994). The banana and sugar industries built at this time continue to supply the region's largest economic export, but tourism is the region's most important source of revenue. Until recently, exports were protected under the Lome Convention, which ensured favorable trading status of the African, Caribbean, and Pacific countries with the European Union. However, the United States' Chiquita Banana Corporation has forced, via the World Trade Organization, new trading arrangements which are now in development.

During the 1800s, the United States' Monroe Doctrine blocked new influence by European countries in the Western Hemisphere and provided the foundation for the future U.S. empire. Later, racist notions of Manifest Destiny justified an expansionist program from the United States. Through three U.S. presidents, Secretary of State William Seward put together an overt plan for "global commercial empire" which tied together the Caribbean and the Pacific (Randall and Mount 1998, 29). Principal to Seward's plans was establishing strategic naval bases in the Caribbean to control commerce.

This plan largely succeeded. The United States has "ship riding" power in most of the region and can interdict ships in foreign sovereign waters in search of drugs (Meeks 2001); in addition, the United States has upheld and instituted dictatorships in the Dominican Republic and Haiti "for most of the twentieth century" (Dupuy 2001, 530), and has undermined every left or progressive (non-neoliberal) government in the entire region since 1980 (Dupuy 2001), with the notable exception of Cuba—but not from a lack of trying.

Other forces of globalization in the region are the drug and the cruise industries. The fact that these industries are not entirely United States based indicates that United States economic power in the Caribbean and the world is not unchallenged. Many of the Caribbean countries cannot adequately police their borders, especially the hundreds of islands and thousands of cays (e.g., the Bahamas has 700 islands and 2,000 cays). Drug traffickers use the area as a cover, often terrorizing citizens who have few options for defense (Griffith 1998). As much as 40 percent of the drugs destined for the United States and Europe go through the Caribbean (Dupuy 2001).

A Note about the Cruise Industry

Tourism has become the most important industry in the region, replacing agriculture as the primary source of revenue (Lewsey, Cid, and Kruse 2004). This has resulted in beach development for affluent visitors, decreased *average* poverty (one-fifth of the jobs now are tourist related, and property values have increased), and disturbed coastal ecosystems. About twelve million tourists visit the region each year, with about eight million of them arriving by ocean (Lewsey, Cid, and Kruse 2004), like the first colonists.

The cruise industry is the fastest-growing major economic sector in the Caribbean. Seventy-one ocean liners (excluding yachts and relatively smaller tourist and charter ships) brought ten million tourists to the region per year as of 1997, and it continues to grow 8 percent per year (Wood 2000). The cruise business relies on multinational labor from over fifty countries on any single ship, but the industry itself is highly concentrated. Three companies dominate two-thirds of the North American market: Royal Caribbean International, Princess, and Carnival. All of these companies use flags of convenience, are based outside the region, and negotiate exemptions from labor and environmental laws (Wood 2000).

Andrew Schulkin (2002) says this industry poses a unique environmental threat. Repeated investigations reveal that cruise ships release into the marine areas they visit several kinds of pollutants, including untreated or inadequately treated sewage water from the passenger toilets, and "scientists now believe that pollution is destroying coral reefs" (Schulkin 2002, 113). According to Oceana, a well-funded United States–based NGO, its campaign to clean up the industry pressured Royal Caribbean to install advanced treatment water systems on its fleet (Oceana 2004).

Tourism also presents some common globalization themes of local disempowerment. In a chapter in another volume within this series on globalization and the environment, *Confronting Environments*, Donald Macleod (2004) describes a village in the Dominican Republic:

> The whole area has been transformed from a natural resource producing goods for food and shelter for local people, to a resource for aesthetic appreciation and recreation consumed by tourists and dominated by hotel chains. The state of local and global power relations in the area determines the socioeconomic lives of indigenous inhabitants, prescribing the range of possibilities in their lives. (32)

Coastal space formerly under common use is now privatized, once abundant fishing has now disappeared apparently from traffic, land use has changed to more hard paths of development like hotels on the beach, and effluents have increased dramatically, all in conjunction with vast economic inequality symbolized by segregated beaches guarded by gunmen to keep island residents out of their former space.

Today, the Caribbean's share of the world economy has dropped from 28 to 15 percent (Dupuy 2001). The bulk of the region's assets are minerals, manufacturing, tourism, and agriculture. Poverty is especially high in the non-English-speaking countries, reaching 45 percent in the Dominican Republic, 65 percent in Guyana, 78 percent in Haiti, and 43 percent in Jamaica (Dupuy 2001). Table 5.1 and figure 5.1 show the degree of foreign direct investment in the region. When the United States is factored out, the region shows the largest rate change in FDI, but in absolute terms it is less than US$2 billion.

By taking out the hubs of capital in each region, we can see that the Caribbean average FDI is in the middle of the other two regions, but highest in terms of rate of change. Table 5.2 shows that the Caribbean is also the middle region for development assistance, and for its number of foreign affiliate corporations hosted. It is important to note that Southeast Asia has the most affiliated corporations by two orders of magnitude, making the Caribbean and the South Pacific more comparable to each other in this way than to Southeast Asia. The Caribbean has the greatest amount of debt to percent of income at 83 percent (table 5.2).

I will now review some of these key ecological trends in the region and follow

Table 5.1. Change in FDI in the Caribbean

Country	Foreign Direct Investment 1988–1990 (inflows, millions of $US)	Foreign Direct Investment 1998–2000 (inflows, millions of $US)
Region	4,151 (average)	18,569 (average)
Region without USA	293 (average)	1,636 (average)
Belize	17	28
Colombia	426	2,224
Costa Rica	129	564
Dominica	116	997
Grenada	—	—
Guatemala	151	353
Guyana	0	54
Haiti	9	18
Honduras	48	206
Jamaica	61	450
Mexico	2,755	12,171
Nicaragua	0	246
Panama	39	850
Trinidad and Tobago	107	671
United States	58,159	255,633
Venezuela	251	4,083

Source: Adapted from World Resource Institute 2003a.

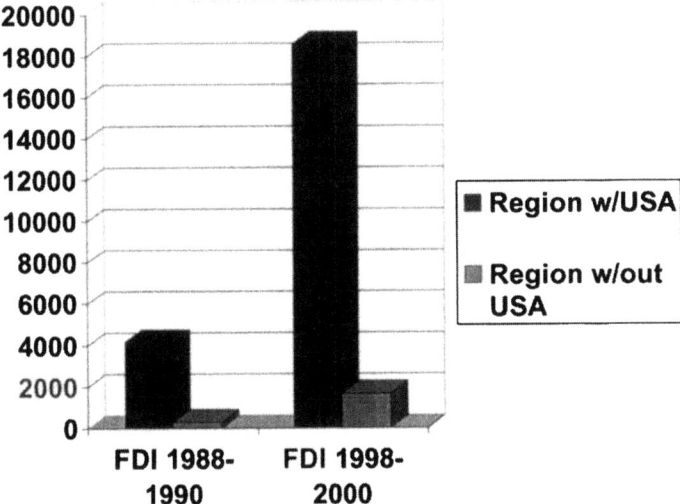

Figure 5.1. Change in Foreign Direct Investment (FDI) in the Caribbean
Source: Adapted from World Resource Institute 2003a.

up with an analysis of the capital and institutions in the region. This analysis enables me to perform the Borgese Test for this incredibly important region.

Ecology

Fisheries

In 1999, Brown and Pomeroy wrote:

> The countries of the Caribbean Community (CARICOM) have a relatively poor record of fisheries management and the need to reform fisheries governance is urgent. Many of the fisheries are fully or overexploited. This is especially true for near shore demersal and coral fish species, conch and lobster, and coastal pelagics on which many of the fishers in the region are dependent for their livelihood. The fishers, most of whom are small scale, are now finding their livelihoods threatened due to resource overexploitation and environmental degradation. In addition, tourism and economic development have caused increased conflicts among various coastal and marine resource users. (549)

In fact, none of the CARICOM countries had fishery management plans, but that has recently changed (Brown and Pomeroy 1999).

Table 5.2. Corporations and Debt in the Caribbean

Country	Parent Corporations	Foreign Affiliate Corporations	External Debt (% of GNI)	Official Development Assistance Receipts (% of GNI)
	1994–2000		1998–2000	
Region (average)	1,231	1,955	83	5.0
Region without USA (average)	153	812	—	—
Belize	—	4	57	3.7
Colombia	302	2,220	38	0.3
Costa Rica	—	111	31	0.1
Dominica	—	92	29	0.8
Guatemala	—	287	25	1.4
Guyana	4	59	228	14.4
Haiti	—	6	30	7.4
Honduras	—	30	108	9.9
Jamaica	—	177	64	0
Mexico	—	8,420	37	0
Nicaragua	—	21	358	31
Suriname		9	—	
Panama	—	279	77	0.2
Trinidad and Tobago	—	65	39	0.2
United States	3,387	19,103	—	—
Venezuela	—	406	42	0.1

Source: Adapted from World Resource Institute 2003a/b.

The region does not have the pelagic fisheries found in the other regions, which may be one reason why major industrial fishing is not found to the same degree in the other areas. However, fish like snappers and groupers are commercially important in the region. Snappers and groupers are high on the fish trophic level (food hierarchy); they are slow to mature; they reproduce late in life; and the fact that they socialize in groups easily caught makes them susceptible to severe exploitation. Subspecies in both fisheries are thought to be overfished or even endangered, but information in the Caribbean region is insufficient to adequately determine their condition (table 5.3; Mendoza and Larez 2004). In the southern Caribbean, catch for the southern red snapper peaked in 1983 and has since declined with varying degrees of effort; similarly, the yellowedge grouper was exploited with increasing degrees of effort until 1991, when it reached its maximum yield, and is now also in decline. Consequently, both species require conservation measures, but these fisheries are now unregulated (Mendoza and Larez 2004).

Table 5.3. Caribbean Fisheries

FAO Designation	FAO Trajectory of Catch	Pressure	Biodiversity	Other Signs about the Fisheries
30% overfished	Leveling out	Moderate pressure from regional fishers which went from 20,000 vessels in 1970 to 60,000 in 1998	Severe fishery biodiversity problem apparent; Atlantic shark populations collapsing 40–99% depending on the species; western Atlantic bluefin tuna down 90% since 1975	Most important fisheries are in decline: fully exploited, overexploited, or even endangered; snapper, grouper, mullet, parrot fish, near shore demersal, coral reef fish, conch, lobster, and coastal pelagics are all almost or fully overfished

Sources: FAO 2002; FAO 2003a; FAO 2003b; Mendoza and Larez 2004; Brown and Pomeroy 1999.

Similar downward trends face parrot fish. As fishing pressure increases in the Caribbean, parrot fish biomass decreases and smaller, younger fish come to dominate communities. Most parrot fish species start life as females and change to males when they reach a large enough size. In a study by Hawkins and Roberts (2004), males of the largest species were virtually nonexistent in island countries with high fishing pressure such as Jamaica and Dominica.[1] On several levels, fisheries in the Caribbean are on a decidedly downward turn toward less biomass and smaller individuals. These changes are having significant related negative effects for the Caribbean marine environment, such as weakened coral reefs. Indeed, in Jamaica, a combination of overfishing and pathogens reduced algae-grazing fish where "coral cover decreased between 1980 and 1993 from 52 to 3%, while the benthic microalgae increased from 4 to 92%" (Holmlund and Hammer 1999, 256). In essence, the coral was replaced with slime because the fish that ate the slime were removed from the area.

Fishery conditions in the Caribbean are very closely related to the reef conditions. Protected reefs appear to have higher levels of primary production, total biological production, and fishery biomass than reef areas which are unprotected and have relatively few constraints (Arias-González et al. 2004). Thus, fishery and coral reef health (not to mention mangrove and sea grass health) rise and fall together—both are currently falling.

Coral Reefs

About 9 percent of the world's reefs are in the Caribbean (Barker 2002). Table 5.4 shows that most Caribbean reefs are moderately or highly threatened. Several pressures, including the cruise ship pollution and fishery declines mentioned above, in addition to ocean warming, are responsible for this threat. It is suggested that if the Caribbean experiences a 1°C increase in surface water temperature for a month or more, the region may experience extensive coral bleaching (Chew 1997). Mass coral bleaching events have occurred during 1983 and 1991.

A study by Gardner and others published in 2003 in the journal *Science* found that the Caribbean reef cover was being reduced by 80 percent across the region with "some degree of synchrony" (958). Importantly, this change operated relatively evenly throughout the region—thus, significant variation did not appear on the country level, indicating that the minimum level of analysis adequate for this issue is the region.

Sea Level Rise and Marine Temperature Change

As table 5.5 shows, the average temperature has risen about 1°C in the last fifty years, and is expected to raise another 2–4 degrees in about another fifty years (Singh 1997). The sea level has risen about 20 cm in the last 100 years, and is expected to rise another 6 cm in the next few decades (Singh 1997).

Caribbean Sea level rise may actually be 8–10 mm/yr, several times above the global mean of 2 mm/yr, but this observation is short of the fifty years needed for time series data on sea level. Consequently, this observation cannot be generalized (Chew 1997).

Table 5.4. Caribbean Coral Reefs

Level of Risk	Low	Medium	High
Proportion of Reefs at Risk (%)	39	32	29
Global Reefs at Risk (%)	42	31	27

Source: Spalding, Ravilious, and Green 2001.

Table 5.5. Caribbean Climate-Related Changes

	Experienced	Projected
Temperature Change	+1°C last 50 yrs	+1.5–4°C
Sea Level Rise	+20 cm last 100 yrs	+20 cm by 2025

Source: Singh 1997; Lewsey, Cid, and Kruse 2004.

Along with sea level rise, beach erosion and saltwater intrusion into the freshwater sources of the area have been a problem. These threats will probably become more severe with or without continued sea level rise, since some islands, such as Trinidad and Tobago, are facing severe subsidence (sinking land) from groundwater and oil extraction (Singh 1997). Subsidence lowers the water table, allowing for saltwater to enter the aquifer the same way sea level rise does, and the region is relatively water scarce (Lewsey, Cid, and Kruse 2004). In addition, warmer ocean water from global warming has increased the severity and quantity of hurricanes in this region (Emanuel 2005; Webster et al. 2005). Thankfully, mangroves and sea grasses have survived these climate changes until now and are expected to survive the threat of climate-related sea changes, though their prognosis from coastal development is a different story.

Regional Political-Economic Conditions

Poverty and Violence

Poverty in all of Latin America rose a small percentage between 1999 and 2002, to encompass nearly one in five residents, and the absolute number of people living in poverty rose between 1990 and 1999 (in 2000 poverty declined) and in 2001 and 2002 to total 221 million people, 98 million of whom were indigent (Economic Commission for Latin America and the Caribbean, 2003).

Levels of affluence in the region truly are mixed. For example, the average per capita income in the region grew by 30 percent between 1975 and 1995, at the same time that the region switched its emphasis from monoculture crops to tourism and its subsidiary services, which now account for 25–35 percent of the region's economy.

Beyond the data used to measure violence, the United States began to wage a discretionary war against Iraq in 2003; the war deposed the despotic regime of Saddam Hussein in a matter of weeks. This war was declared over, but as of this writing it is still being fought against guerrilla insurgents through a United States–appointed Iraqi government and United States–led occupying forces. There was also bloodshed in Haiti in 2004, during the period when the United States mysteriously flew the democratically elected president Jean-Bertrand Aristide out of the country during a violent rebellion. Beyond that, there has been violence *from* the region in two conflicts in 2000 and 2001, which involved one war and one intermediate armed conflict by the United States with Al Qaeda and the Taliban as it fought in Afghanistan. Near the Caribbean region, and affecting the region, there have been since 1965 several armed conflicts in Colombia between the government, funded heavily by the United States, and the Fuerzas Armadas Revoluci-

Table 5.6. Poverty and Government Expenditure in the Caribbean

Country	% Population Living on Less than $1/day	% Population Living on Less than $2/day	Military Expenditure (% GDP) 2000	Health Care Expenditure (% GDP)	Education Expenditure (% GDP)
Region	14	33	2	4	4.7
Region without USA	—	—	1	3.7	4.5
Belize	—	—	—	2.3	—
Colombia	19.7	36	2.3	5.2	—
Costa Rica	12.6	26	—	5.2	6.1
Dominica	3.2	16	—	1.9	—
Guatemala	10	33.8	0.8	2.1	2
Guyana	—	—	—	4.5	—
Haiti	—	—	—	1.4	—
Honduras	24.3	45.1	—	3.9	4
Jamaica	3.2	25.2	—	3.1	6.4
Mexico	15.9	37.7	0.5	2.6	—
Nicaragua	—	—	1.1	8.5	4.2
Panama	14	29	—	4.9	—
Trinidad and Tobago	12.4	39	—	2.5	—
United States	—	—	3.1	5.8	5
Venezuela	23	47	1.2	2.6	—

Source: Adapted from World Resource Institute 2003a/b.

onaris Colombianas (Revolutionary Armed Forces of Colombia). These conflicts have ranged in scale from minor armed conflict to outright war (causing over one thousand battle-related deaths) (Eriksson, Wallensteen, and Sollenberg 2003). This war is fought partially over the drug trade and partially over oil resources. Since a large amount of the drug trade passes through the Caribbean, this conflict relates to the basin and will continue to affect it, but it is not technically included here since I am not including Colombia within the LAC.

Fisheries

From 1988, the Caribbean region went from catching 87,000 tons to 172,000 tons of fish from its marine areas, half of which come from Cuba and St. Vincent/Grenadines in that order. This means that the region went from making up 0.1 percent of the world marine catch to 0.2 percent (FAO 2002), confirming the reports that this is not a major international source for the fishing market or industry. However, the fisheries are still being depleted and simplified as described above. In trade, the region imported 67,000 tons in 1988 and 102,000 tons in 2000; it exported 82,000 tons in 1988 and 211,000 tons in 2000. Overall, production and trade in fish doubled over this twelve-year period (FAO 2002), but they remain low in absolute terms compared to Southeast Asia (as well as other regions not included here, like East Asia and Europe).

In the 1980s the Organization for Eastern Caribbean States (OECS) and the CARICOM Fisheries Resource Assessment and Management Program harmonized their fishery laws (Brown and Pomeroy 1999). Unlike the South Pacific and Southeast Asia, fishing in the Caribbean is mostly small scale and locally based. The FAO includes the number of fishers from all of North and Central America, but even with this aggregation, this region has only about 4.5 percent of the word's total fishing fleet, which grew from about 20,000 boats in 1970 to about 60,000 vessels in 1998 (FAO 2002). However, the fisheries in the Caribbean are largely centrally controlled by the State. This is a legacy of slave-based political economies, which dictated at the plantation level which slaves were permitted to fish (Brown and Pomeroy 1999). Thus, local "comanagement" was not prevalent, as it was in the South Pacific.

Coastal Development Projects

Within the wider Caribbean region, there is general agreement that coastal marine systems are changing for the worse. The ultimate causes are population growth and human-induced changes both in the coastal zone and in the upstream watersheds. Mangroves are removed for lumber, charcoal, aquaculture, and land reclamation. Seagrasses are being dredged for harbors or for beaches in hotel development and deforestation inland is leading to increased runoff and sedimentation. Nutrients from sewage and agricultural runoff increase eutrophication, and coastal fish stocks are being depleted. (Linton and Warner, 2003, 264)

Linton and Warner document marked pollution coming from coastal-development projects, including sewage systems, because only about 10 percent of the Caribbean population has access to treated sewer systems. Agricultural projects and shipping are also factors. In addition, Antigua, Barbados, and the Bahamas have seen all of their native forests eliminated as a result of urban and coastal development, which has in turn concentrated polluted runoff. Most of the urban areas are on the coast, and many of these are in coastal floodplains which are particularly vulnerable to increasing sea levels and storms. This development has replaced over 1,600 sq. km of mangroves in the Bahamas, which constituted over 50 percent of the area's holdings (Lewsey, Cid, and Kruse 2004).

Lewsey, Cid, and Kruse (2004) note that the region is particularly sensitive to coastal changes, such as climate change, and that coastal development has been neither planned nor ecologically benign. Furthermore, the institutions that are in place to deal with current and foreseeable problems are not sufficient:

> Few areas in the world are more vulnerable to climatic variability than the low-lying island states in the Eastern Caribbean Basin. While their small land masses leave them vulnerable to hurricanes and tropical storms, that vulnerability has been exacerbated because of human activities—intensive land development, high population density in coastal zones and poorly developed coastal infrastructure are complicated by the impacts of tourism-based industries, limited human and cash reserves, and a lack of trained personnel who can address the impacts of climate variability. (393)

The Caribbean coastal zone is demonstrating mixed but disconcerting trends. There are efforts at protecting the coastal zone through an increase in MPAs, but these efforts seem piecemeal and notably outraced by pressures to build and pollute these zones.

Cuba, in facing serious ocean-related decline, provides a counterpoint. For example, Cuba provides the most pressure on fisheries in the region and has seen a 95 percent decrease of landings in groupers and an 88 percent decrease in landings of mullet (Benchley 2002, online). That being said, the director of the Center of Marine Studies at the University of Havana noted in *National Geographic* that "Because the government controls all levels of activity . . . implementation of order is easier than in other Caribbean countries. There is not much violation of our laws. As a result our marine environment is in better condition than elsewhere" (quoted in Benchley 2002). Thus, Cuba has been ostracized from the rest of the CARICOM, and lags some "40 years behind in terms of massive tourism development and the concomitant destruction of marine life and habitat" (Benchley 2002).

This situation probably results from a lack of capital to build the physical structures of industry and tourism. Another influence may be knowledge production; this is one of the strongest assets in Cuba, and one of the greatest needs for the Caribbean sustainability in general. For example, Cuba has more than 9,000 employees, including 350 PhD's, working in its Ministry of Science, Environment and Technology formed in 1994 (Benchley 2002).

As of 1997, 45 percent of the Caribbean was under protected status. Placing this area under protection marked a way to address sustainability issues through planning, with the aim of preserving biodiversity as the premier conservation goal (Zimmerer and Carter 2001; Meeks 2001). This coincides with the "mushrooming" of local and national environmental NGOs which are collaborating and networking with international groups like the World Wide Fund for Nature (WWF) or the World Conservation Union (IUCN) and the Nature Conservancy (Zimmerer and Carter 2001; Meeks 2001). Thus, as neoliberal markets penetrate the region and place a heavy emphasis on resource extraction for economic growth, there is also a connectivity (which may be generated by the very same liberalism) that is occurring between civil-society groups. However, note that the stronger NGOs are not local, but are from the hubs of capital themselves—the United States or Europe. This demonstrates that the same opportunities for crossing regions does not exist for Caribbean local civil society unless these groups partner with the large groups like the WWF.

In addition to all of these issues, global warming will have a negative effect on the region's coastal zone. Imminent climate change has generated a common cause within the small island states, including those in the Caribbean where adverse impacts will be felt by these relatively innocent agents of atmospheric change. Note from tables 3.3 and 3.4 that there are no independent Caribbean states that make the list of top twenty-five countries by emissions in absolute or per capita terms except Trinidad and Tobago, which has an oil industry. As in the South Pacific and in most Southeast Asian cases, the countries in this region are feeling the effects of this problem, but are not enjoying the benefits that come with large-scale carbon emissions. Singh (1997) writes:

> It is a widely held belief among scientists, environmental advocacy groups and politicians in the small island developing states (SIDS) of the Caribbean, that global warming and sea level rise are being imposed upon them by the developed world. Furthermore, because of economic limitations, they feel helpless in their efforts to develop policy options for dealing with these regional changes. Also, because of their relative sizes and economies, they do not fully perceive the need to reduce GHG emissions and to curtail deforestation. These issues therefore pose serious threats to environmental sustainability in the Caribbean, and on a longer time-frame, to international security. (95)

Thus, specific regional institutions, the States, and civil-social pressures seem to be ineffective, at least as of yet, to significantly, if at all, mitigate global-level climate forcing. However, this does not mean that the region is without agency; it means that the agency it has exerted until now has been met with undeniable brute and economic force from much larger actors. Dujan (2002) notes that neoliberal globalization has meant an ideological preference for "free trade" principles which have undermined a great deal of Caribbean efforts, including defending preferential access to markets for their agricultural products like bananas. However, she describes a form of effective resistance found in St. Lucia, where traditional land tenure practices became embedded in enforceable property rights that survived neoliberal tests and protected locals with some stability. She believes this outcome is theoretically important and applicable to globalization studies because it shows how to use neoliberal values for local protections and resistance. This resistance is important because the otherwise compromised position of the small states has allowed neoliberal trade and lending policies that have led to job loss, cuts in health care, education, and social services, and increased costs of living in the Caribbean (Dujan 2002).

This kind of effort could potentially be used in coral reef and fishery protections but not against climate forcing. Therefore, among SIDS, there is continued lobbying of the industrial countries to change their greenhouse pollution practices, but the real focus is now on "adaptation" like sea walls, agricultural plans, and human relocation and other coastal-zone modifications.

Finally, the issues of shrimp farming are applicable to the LAC, particularly in Honduras, where the shrimp-farming industry has boomed since the 1980s thanks to private foreign investment as well as money from the United States Agency for International Development, other U.S. agriculture-based federal loans, and the Inter-American Development Bank (Lemay 1998; Stonich and Bort 1997; DeWalt, Vergne, and Hardin 1996).

Shrimp mariculture has resulted in systematic violence and exclusion of subsistence fishers from coastal commons, in addition to abhorrent levels of pollution, according to Stonich and Bort (1997). They also note that this industry is indeed a global phenomenon given that 99 percent of the shrimp are grown in poor countries in the LAC and Southeast Asian regions and consumed in affluent nations in North America, Europe, and East Asia (most importantly Japan). Further, they note that there is a "high degree of foreign capital and technology necessary to maintain the industry in its present model," and that the current policies that focus on export earnings will expose the region to a "continuing social crisis, and an increasingly competitive global production system" (Stonich and Bort 1997, Ebscohost online).

It is interesting that Stonich and Bort believe that NGOs have had more success in fighting the practice of shrimp farming in Honduras than in the Philippines, as evidenced in the work of groups like the Committee for the Defense and Development of the Flora and Fauna of the Gulf of Fonseca. The authors note that members of this group "and others who have resisted the expansion of the industry have been flagrantly harassed, lost their jobs, been imprisoned, received death threats, and been murdered" (Stonich and Bort 1997, Ebscohost online). This practice is reducing biodiversity as it replaces diverse mangrove estuaries and salt flats with the monoculture of shrimp in Ecuador, Mexico, Panama, Guatemala, El Salvador, and of course Honduras. However, this change has not yet been extensively found in the Caribbean basin. Nonetheless, Stonich and Bort firmly state that this practice will become entrenched in the region because power now resides in a "new generation of leaders committed to free-market economies and reduced regulations" (Stonich and Bort 1997). Further, they believe it is "conceivable that significant portions of both the Pacific and Caribbean coasts of Central America will be transformed into shrimp farms in the near future" (Stonich and Bort 1997).

Knowledge Production

Given the overwhelming influence of European colonization on other practices, it would be surprising to find that knowledge production has occurred outside of this influence. Indeed, the extent of this influence is disputed, but it is not disputed that European notions of identity, culture, science, place, and nature—in a sense, European human ontology—have had a strong enough effect that many authors call for a decolonization of philosophy and the Caribbean mind (see, for example, Benn 1987). Paget Henry (1998) notes that the continued colonization of knowledge is part of the "savage capitalism" and neoliberalism that the region is experiencing, and is a critical aspect to the loss of sovereignty not just in the Caribbean, but in the Third World generally. Part of this loss of sovereignty is found in the loss of legitimate alternative academic inquiry, which faced a "virtual damper on what was considered the acceptable field for research" by neoliberal ideologies that have been intolerant of leftist alternatives (Meeks 2001, xiv).

Thankfully, at the same time there are strains of "new Caribbean thought," which seems to continuously refer to liberalism, modernity, and European colonial influences but is also uniquely Caribbean—perhaps even a renaissance in the contemporary period which includes cutting social criticism, art, economics, and a rich interdisciplinary set of voices (Meeks and Lindahl 2001). In addition, there has always been a distinctive ecological Caribbean source of knowledge building, for example, in botany. In his book *States of Nature*, historian Stuart McCook describes agricultural and biological research in five Caribbean countries and finds that science was both local and foreign driven. He writes,

> Given the large role that the Unites States assumed in the Caribbean during this period, it might be tempting to tell this story in terms of U.S. imperialism. After all, most centers for plant research in the Spanish

Caribbean were modeled after institutions in the United States. Moreover, many U.S. scientists worked at these institutions, and many Latin American scientists at the same institutions had been trained in the United States. The United States was the primary market for most of the crops they studied. But even so, it is impossible to explain the growth of science in terms of the United States alone. (McCook 2002, 4–5)

The antecedent focus of this mixed science, according to McCook, was always toward pragmatic agronomy which was coordinated with state building so that the production of biological science and agronomy became closely tied to the existence and development of the respective nation-state. In this vein, he quotes Cuban sugar grower Jose Manuel Casanova, "*Sin azúcar, no hay país,*" or "without sugar there is no nation" (McCook 2002, 1). The result of some of this science was to overemphasize export conditions and modernist agronomy over the long-term ecological needs of the area. Now nation building and science seem to be continuing this trend with an eye toward ecotourism.

The production of knowledge, particularly as it relates to environmental issues, is also important in the region and is conducted in part through NGOs. One such regional NGO is the Caribbean Conservation Corporation, which is based in Barbados and has member organizations throughout the Caribbean. The organization works to enhance cooperation in the region; aid and establish marine protected areas; and build scientific and political intraregional knowledge. It has a nonwhaling program and an ecotourism concern. Because it is an organization that is involved across regions, particularly in partnership with other small island areas such as the South Pacific, it can be seen as a global organization. Apparently much of the organization's contact is accomplished by travel, but most of it is through electronic communications, and includes environmental news and research connections, such as on its website at www.ccanet.net. The Caribbean Natural Resource Institute described below is another example of this kind of regional and transregional networking and knowledge building.

Regional Institutions and Civil Society

Nation-states in the region have organized around environmental concerns at the same time they are organizing in favor of a transnational tourism business. In the last twenty years, the wider Caribbean has instituted important international environmental treaties, an accomplishment that shows "mechanisms for cooperation that are positive models for the rest of the world" (Barker 2002, 83). Thus, institutions are including environment in their rhetoric, but their capacity is weak.

The lack of "awareness, strong institutions, and human resources continue to impair the sustainable use of natural resources of the region" (UNEP quoted in Barker 2002, 82). However, I do not think we should view the confluence of weak environmental governance and a rising neoliberal order as accidental.

Macleod (2004) demonstrates in the Dominican Republic village case discussed above that the interests of the States and transnational firms are "clear winners." This is representative of the order being formed at the regional level generally. Civil society is not passively lying down—in the above case, citizens protested to gain some concessions, such as partial access to some of the beach. But, as with other civil-society struggles found in this book, the citizens' ability to reach across the ocean is not as strong as that of the hotel firms, and they are stuck fighting a globalizing event at the local level, which may be structurally ineffective. Macleod poignantly describes this exclusion and control of the coastal zone, for example, fishing grounds, freshwater, land use, and sea access, as a map of the State and corporate power since the representation of nature and space by these two entities is produced in contrast to that of the locals. In this way the institutional settings of firms and State are sympathetic to one another, but relatively unaccountable to local civil society or its ecology.

The status of women in particular is of major concern. Oma Panday, wife of the prime minister of Trinidad and Tobago, launched the Caribbean Network of Women Producers in 1999, at which time she said:

> There is information that women produce between 60 to 80 percent of basic agricultural foodstuff in the Caribbean. Yet in spite of their major role as agricultural producers, women own less than 2% of the land. One reason [for this] is that it is very difficult for rural women to get access to credit and that very many women, particularly young women, are severely handicapped by limited education. (Women's International News Network 1999, 22)

Indeed, research over a ten-year period shows that women in the region suffer more poverty than men, even though women in the area are more educated in general. Most indigent households are headed by women, as are most single-parent poor families (United Nations Economic and Social Commission for Latin America and the Caribbean, 2003).

These findings indicate a vast discrepancy in women's political power which is subject to a steep hierarchy which in itself is not sustainable. In addition, it is apparent that minority groups within the region, such as the Rastafarians and indigenous Carib peoples, are also held in marginal positions. This reflects a non-egalitarian society, that is, a steep social hierarchy. These are important issues which deserve more elaboration, but for now it will have to suffice to conclude that gender and ethnic oppression continue throughout the region in ways that probably keep problems and solutions from that level elusive.

There are several regional governing institutions in the Caribbean basin. Regional integration per se is not new to the area since the first attempt in the 1950s by the West Indian Federation, which was developed in order for Britain to have more efficient control over the area without having to deal with ten different colonies. However, the kind of regional organization needed to deal independently with global pressures of commerce and changing ecology is new. This regional identity is fragile. Despite a deep set of differences among them, the Caribbean countries have a history of cooperation with one another, but in the independent states there is also a strong theme of protecting newly acquired sovereignty and relative autonomy from colonial powers (Knight and Persaud 2001; Griffith 1998).

The most prominent formal regional institution is CARICOM, which is comparable to ASEAN and the South Pacific Community institutions. It encompasses the Caribbean Common Market and includes the fifteen countries mentioned at the beginning of this chapter. This institution was initially organized in 1973 for economic growth by Barbados, Guyana, Jamaica, and Trinidad and Tobago—which are also the four countries in the region which have Structural Adjustment Loans from the International Monetary Fund (IMF). Thus, the regional politics have formed in relation to economic globalization, but in favor of them, not in resistance to them as Hettne (1999) expects from regional groups.

The effects of these loans seem to have been mostly negative. The austerity adjustments have forced decreases in social programs, and the purchasing power of the citizens has declined through devalued currency adjustment. Economic growth has not been realized and has even added to the disenfranchisement of the poor, particularly in Jamaica (Elu 2000; Meeks 2001).

Nonetheless, many consider CARICOM one of the most important successes in the region. An officer for Caribbean Latin American Action, a United States–based lobbying group, noted that "Pretty much for the first twenty years, Caricom was seen as probably one of the world's best examples of regional integration schemes" (Lewis quoted in Luxner 1999, 56).

A diplomat from Trinidad who has been with CARICOM since its early days notes,

> Our goal is free movement, not only of goods, but of services, capital, and skilled people across the entire Caribbean Community—something close

to what the European Union has done. . . . As small individual units, we are not viable politically or economically. We are much more viable as a group. (Lewis quoted in Luxner 1999, 56)

As such, CARICOM was created out of desperation in the 1980s to deal with deteriorating economic conditions like balance of payment deficits, increased cost of importing food, and low productivity (Elu 2000, Infotrac online). Thus, the regional cooperation found in CARICOM is aimed at increasing economies of scale. Nontourist economies of the Caribbean are based directly in natural resources through agriculture, mining, or fishing and logging, all of which have strict limits on scale and rate of exploitation. Also, payment for these goods is graded lower than other types of trade, which means that the area has been structuring itself toward a depletion of resources for minimal, unequal trade set by the precedent of early colonialism.

The Organisation of Eastern Caribbean States, another regional organization, includes Antigua and Barbuda, the Commonwealth of Dominica, Grenada, Montserrat, St. Kitts and Nevis, St. Lucia and St. Vincent, and the Grenadines. This organization articulates its mission as:

> to be a major regional *institution contributing to the sustainable development of the OECS Member States* by assisting them to maximise the benefits from their collective space, by facilitating their intelligent integration with the global economy; by contributing to policy and program formulation and execution in respect of regional and international issues, and by facilitation of bilateral and multilateral co-operation. (OECS 2004, emphasis added)

This particular Caribbean regional institution states sustainability up front as a primary mission. However, much of the organization's efforts are geared toward brute economic development. Units of the organization include a legal department for the harmonization of laws and departments that work on education and human resource development; export development; environment and sustainability; pharmaceutical development; social development; and national sports development. The environmental sustainability department generates development reports, provides technical assistance, and highlights opportunities for economic development with environmental benefits. The department argues that its mission is guided by the St. George's Declaration of Principles for Environmental Sustainability, which among other things argues that environmental quality is essential to the security and well-being of the Eastern Caribbean people. The eradication of poverty, the inclusion of civil society and the private sector, adaptation to climate change, and the protection of cultural heritage and biological diversity are specific intentions of the Eastern Caribbean nations in this organization. The program has many partners, from United Nations agencies to the United States Department of State to regional Caribbean actors such as the University of West Indies, discussed below, and CARICOM (OECS 2004).

Comparison to the Borgese Test of Sustainability

There are clear indications that the Caribbean faces incredible threats to its social and ecological future. The ecological threat is made much worse by poverty, which can reasonably be traced to the region's colonial history as a base for the slave trade and U.S. and European imperialism. This legacy has institutionalized steep hierarchical power relations in the region, richly illustrated by the leagues of middle-class tourists who come to the islands on tours the people of the islands typically cannot afford. This hierarchy may be informed to some degree by the precolonial social hierarchies, as well as those hierarchies which were reproduced by imperialism.

The ecological threat demonstrated by the indicators looks like this: major fisheries are declining and becoming less diverse, coral reefs are mostly moderately or severely threatened with heavy mortality in the coming few decades, and regional sea temperatures are on the rise along with sea level rise. The latter is currently a very serious local threat to the small island developing states in the region. None of these biophysical indicators are improving, and the future of the Caribbean looks very difficult and even more insecure than it is now. Globalization through the concentration and mobility of capital appears to be a significant factor in aggravating—if not causing—the conditions of reefs threatened with pollution and some of the coastal-zone projects. Local governance to protect fisheries, reefs, and of course climate-related ocean changes has been mostly ineffective. This seems to indict the country-level institutions, as well as the previous waves of globalization that these rest upon.

Violence is often found in the region. Haiti continues to suffer waves of violence, Cuban governance is oppressive in many ways, and drug trafficking plagues the region. All of these issues fail the Borgese Test and indicate severe problems for future sustainability.

Furthermore, there are several examples of steep sociopolitical hierarchies within the region from the global to the family level. Land use planning, one of the great unsustainable features of the region, is top-down and inherited from colonial predecessor governments. There is also the problem of ignoring local needs and vulnerabilities to oncoming climate-related changes, pollution, and other ecosystem issues (Lewsey, Cid, and Kruse 2004). However, a recent study has shown there is a glimmer of hope for budding comanagement of select fisheries. Comanagement at the community level is not common or developing strongly, but it has worked in the Caribbean in the past and a regionally based and focused NGO in the area, Caribbean Natural Resources Institute (CANARI), advocates for it (Warner 1997).

CANARI supports direct citizen research, such as that on near-shore coral reefs in St. Lucia. This report found overexploitation of the reefs and other traditional resource bases, particularly seamoss, snappers and groupers, and conch. Snappers and groupers were identified as threatened primarily by regional Venezuelan fishers, and conchs were overexploited by Martinique fishers (Renard, Smith, and Krishnarayan 2000). Importantly, the citizen research was designed to consider the needs of harvesters at the village level, but used a peer-reviewed methodology for the study and took into consideration previous literature on the assessment of coral reefs. Therefore, the study was ecologically oriented, politically sensitive, and methodologically rigorous. This local-regional effort to generate knowledge specific to the region in order to empower local people is a hopeful sign, but it is an exception to the rule for the Caribbean.

Synthesizing Ecological and Political Structures to Their Global Structure

In sum, all of the ecological and most of the sociopolitical trends are in dangerous decline. Fisheries and biodiversity are declining, the majority of reefs is extensively threatened, and the region is warming faster than the norm. The temperature change will bring with it more intense storms just as the protective reefs and mangroves are being cleared for unsustainable coastal development. Governance to stem these factors has been compromised both by the history of colonialism and by the spread of neoliberal policies which have increased access to more corporations; at the same time, poverty and deprivation in the region are staggering and not being substantially addressed. Violence is moderate compared to Southeast Asia.

Transregional tourism is the major globalization force in this region. It is also behind some of the depletion of fisheries and decline in coral reefs, and is extensively involved in converting the coastal zone into elite zones of consumption. These phenomena have produced similar effects as the forces of globalization, as in Southeast Asia, but on a less intense scale. For example, there has been a loss of common resource use as the region becomes converted into private areas for Northern visitors. It has provided some minor and moderate connections for

NGOs in the region, but to the extent these NGOs are transregional, they are also working *with* the tourism industry, for example, in habitat preservation. Consequently, regional and global NGOs are not a direct source of resistance to globalization from the region. They are negotiating within neoliberal globalization, just as the knowledge production in the region is negotiated and largely foreign based. To what extent are these facets of civil society responsible for some of the Caribbean's problems of sustainability? I suspect these groups are faced with very tough choices that are not new to globalization. If local NGOs or academics pitch actual resistance to the TNCs or the Caribbean states that structure the neoliberal political economy, their influence in the larger scheme is reduced. If they negotiate details of the globalization, they have more influence and credibility within neoliberal policy making, but are forced to accept the basic structural designs of power relations between TNC, Northern governments, and financial international organizations. Resistance at the local level, such as in the Gulf of Fonseca, is important, but it also appears to be perpetual, mostly heavily overpowered, and mostly defensive.

Again, foreign fishing fleets are not the probable reason for grouper and snapper declines, and this is a cautionary example of overemphasizing TNCs. But the depletion is not unrelated to the first wave of globalization and the institutional fishing practices inherited from slavery or the tourism in the current third wave. Also, the IMF structural adjustments have limited the role of government in at least four countries where these loans influence public policy. Overall, the intensity of capital relationship is moderate here compared to in the other regions, just as the relative changes to its ecopolitical changes are also (comparatively) moderate.

Notes

1. Thanks to Julie Hawkins for her assistance in interpreting this research.

Sustainability in Southeast Asia 6

> *The economy of Southeast Asia, a region widely considered to be crucial in terms of economic and environmental development, is growing at a rapid pace after recovering from a slump in the early 1990s. Together with rapid population growth, this has resulted in rising pressures on natural resources in the region, e.g., the destruction of vast stretches of tropical rain forest, loss of biodiversity, and wastes and emissions in urban conglomerates. (Grubuhel et al. 2003, 53)*

SOUTHEAST ASIA IS AMONG the most biologically important and most threatened marine areas in the world (World Conservation Union 2004). The Philippines and Indonesia form the core of this biotic treasure, called the "Fertile Triangle," which harbors "the richest, most diverse marine biota in the world—more than 2,500 fish species of 165 families" (Safina 1998, 308). This alone means that the region's sustainability is tied in a holistic way to that of the rest of the World Ocean as a source of genetic production and wealth. Yet within and outside the region, capital interests primarily from the United States, China, and Japan have made blunt economic growth through marine resource exportation the preeminent goal, leaving poverty reduction and ecological health secondary. The effect is that the biodiversity is impoverished while specific firms and local elites run away with the store, without a proportional feedback indicating this damage.

Southeast Asia has only recently created a formal regional identity. This region's borders blend with other regions such as East Asia, South Asia, and the South Pacific. Southeast Asia has supported a rich array of diverse cultures, geographies, and species in its tropical climate (Sponsel 2000). There are close to a thousand languages, of which 672 are in Indonesia, 167 in the Philippines, 146 in Malaysia, 101 in Burma, and 90 in Laos (Sponsel 2000). However, the identities and ecologies of Southeast Asia are jointly being eroded (Sponsel 2000). This is a pretty good indication in itself that the region is not heading toward sustainability, especially given that institutional rules for some of these cultures have historically been more sustainable.

Southeast Asia is made up of the mainland with Burma, peninsular Malaysia, Thailand, Cambodia, and Vietnam (I add eastern India, Bangladesh, and southern China), and the insular area which supports the area's islands of Indonesia, Singapore, Brunei, the Philippines, and the island area of Melanesia (I add Sri Lanka and now East Timor) (Sponsel 2000). In the insular region, there are thousands of islands, most of which have tropical reefs and rich fisheries. For example, Indonesia has over 17,000 islands (Burke, Selig, and Spalding 2002). This tropical region contains a bulk of the world's mangroves, a third of the world's coral reefs, and one of the world's most productive fisheries. Fish make up around 40 percent or more of the region's protein source.

The region is home to many endemic species (unique to that area). Indonesia alone has over 20,000 species of plants, and almost 3,000 mammals, birds, amphibians, and reptiles (Sponsel 2000). This chapter discusses these changes and describes how some of these forces of globalization appear to be threatening sustainability in the region.

Historical Connections to Globalization

Before European contact, people of the Southeast Asian region had established connections and trade routes from Africa to China, which was the largest market for Southeast Asian–traded goods (Sponsel 2000). This action involved resource extraction, but "the sheer scale and intensity of resource exploitation marked out the colonial era from earlier periods" (Bryant 1998, 31). Europeans first came to the region in the mid-1500s when the Dutch established the Dutch East India Company.

Robert McMahon (1999) notes, however, that Western control over the region did not emerge until the nineteenth century, and the description of the region as "Southeast Asia" is a colonial remnant which did not become popular until after World War II. Before that period, the region (which he believes did not have a common regional identity until recently) was described mostly by the major powers that flank it. For example, the French imposed on Vietnam the term "Indochina," which reflected India and China. I call the region "Southeast Asia" here for convenience and because the region now self-identifies as such.

The entire region except Thailand was under colonial rule at some point. For much of this time the region was only nominally under the influence of European powers (McMahon 1999). However, from 1880 to 1920, European control, led by Britain, became highly centralized, coercive, and violent—obliterating any resistance led by indigenous citizens. The kingdoms of Aceh, Bali, and Sulu were utterly destroyed in this process (McMahon 1999). The geography of the region was ordered as such: Great Britain controlled the Malay states, Burma, and of course India (including present-day Bangladesh). The French controlled Vietnam and Laos. After fighting the Spanish and Filipinos, the United States controlled the Philippines as of 1898. The Dutch controlled Bali and the East Indies. Among the changes instituted by the centralized bureaucracies of these colonial powers, land management, economic structures, and development policies were dramatically altered. In studying this period of "high colonialism," McMahon notes that several trends that are now seen in the region were initiated under this oppression.

> This period witnessed burgeoning urbanization, incipient industrialization, massive immigration, soaring birth rates, and a tremendous spurt in productive economic activity. In order to increase the profitability of their dependencies, the Western colonizers encouraged the systematic development of large-scale economic enterprises geared to the export market. (4)

As today, "Those enterprises were generally capital-intensive, privately financed, and corporately managed" (McMahon 1999). Southeast Asian indigenous peoples negotiated a great deal their own position and governance, but this agency existed in a "highly constricted environment" that was "manipulated and controlled by those who possessed a monopoly on the coercive powers of the state—namely, the British, French, Dutch, and American colonial authorities" (McMahon 1999, 5).

Currently, the amount of FDI, number of corporations, and external debt indicate very intense globally rooted economic change over time. The region has the highest average FDI, and the highest degree of change when the hub is included. No other region in this study came even close to this kind of increase in change in capital penetration. The region also has, by two orders of magnitude, more affiliate corporations within its borders than either the Caribbean or South Pacific. Clearly these indicate a very high degree of economic expansion in the region, and much of this expansion came from abroad. The Asian economic crisis of the late 1990s, which occurred alongside the spread of neoliberal policies by the World Bank and the Asian Development Bank, demonstrated this foreign role. All of this occurred at the end of the Cold War, which ended the ideological and material support from the Soviet Union. This allowed for the United States to become unquestionably hegemonic in the region, pushing its interests through the international financial institutions like the World Bank and IMF (Beeson

2004). There is an irrefutable coordination of neoliberalism and the expansion of capital in Southeast Asia (see tables 6.1 and 6.2).

Ecology

Fisheries

Table 6.3 notes that fishing pressure in this region has exploded since the 1970s. This region has some of the most productive fishing countries in the world, estimated at around 15 million tons a year, minus China. However, the physical region of the Eastern Indian Ocean FAO statistical area produces only 4.7 million tons a year (FAO 2002), indicating that these countries must take at least two-thirds of their catch from other regions. Aquaculture is also a primary fishery source. Compared to the productivity of other countries within the region, China stands out. In 1980, it reported around 5 mmt of total fish catch, and by mid-2000, it was raking in over 40 mmt, or close to one-third of the world's fish catch, 17 mmt—or about one-fifth—of which was from the ocean (FAO 2002). Thus, in this phase (1980 to the present) of globalization, China's pressure on fisheries exponentially increased along with global capital expansion that went from well under 100,000 decked fishing vessels to almost 500,000 (FAO 2002).

Table 6.1. Change in FDI in Southeast Asia 1988–2000

Country	Foreign Direct Investment 1988–1990 (inflows, millions of $US)	Foreign Direct Investment 1998–2000 (inflows, millions of $US)
Region	12,480 (average)	63,239 (average)
Region without China	9,122 (average)	22,938 (average)
Bangladesh	2	217
Cambodia*	0	130
China	3,358	40,301
India	168	2,373
Indonesia*	784	2,550
Laos*	4	66
Malaysia*	1,573	1,792
Myanmar*	56	274
Philippines*	676	1,630
Singapore*	4,039	6,634
Sri Lanka	36	181
Thailand*	1,775	5,631
Vietnam*	9	1,460

Source: Adapted from World Resource Institute World Bank Data 2003a.
Note: Data not available for Brunei* and Burma.
*ASEAN member.

Table 6.2. Corporations, Debt, and Assistance in Southeast Asia

Country	Known Parent Corporations	Foreign Affiliate Corporations	External Debt (% of GNI)	Official Development Assistance Receipts (% of GNI)
	1994–2000		1998–2000	
Region	879	428,023	68.2	3.74
Region without China	500	63,678	73.4 (average)	4.06 (average)
Bangladesh	—	161	35	2.7
Cambodia*	—	598	75	11.3
China	379	364,345	16	0.2
India	187	1,416	23	0.4
Indonesia*	313	2,241	118	1.4
Laos*	—	669	162	20.1
Malaysia*	—	15,567	53	0.2
Myanmar*	—	299	—	—
Philippines*	—	14,802	64	0.8
Singapore*	—	24,114	—	0
Sri Lanka	—	305	59	2.2
Thailand*	—	2,721	76	0.7
Vietnam*	—	1,544	69	4.9

Source: Adapted from World Resource Institute World Bank Data 2003a/b.
Note: Data not available for Brunei* and Burma.
*ASEAN member.

All together without adjusting for overinflated Chinese reports, the region produced about 32 mmt of the world's 82.5 mmt of wild marine fish catch in 2001, or almost 40 percent (FAO 2002). Since about 76 percent of the world's trade (in value) goes to the United States, Japan, and the European Union, this region is a major global producer for the world fish markets (FAO 2002).

Some fisheries are threatened by the massive destruction of coral reefs. For example, Philippine overfishing has caused both a drop in fishery diversity and quantity and a major decline in the coral:

> Philippine coral reef area, the second largest in Southeast Asia, is estimated at 26,000 km² and holds an extraordinary diversity of species. Scientists have identified 915 reef fish species and more than 400 scleractinian coral species, 12 of which are endemic. A large coastal population, rapid population growth of about 2.3 percent per year, high poverty rates, and fisher overcapacity have resulted in major overexploitation of Philippine reef fisheries. Demersal fish stocks are biologically and eco-

116 CHAPTER 6

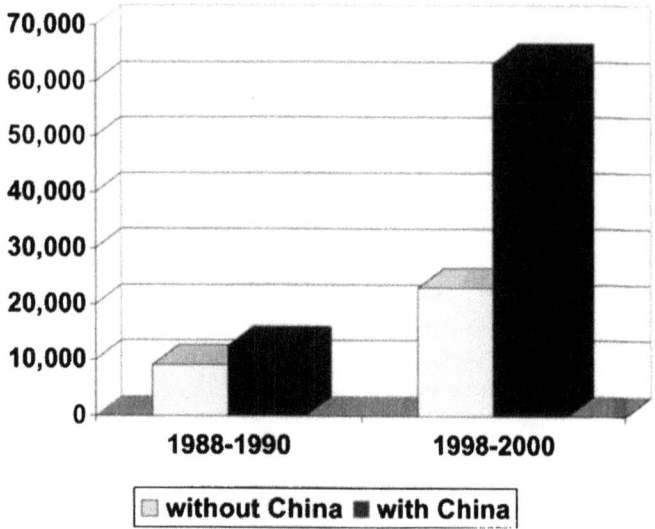

Figure 6.1. Change in Foreign Direct Investment (FDI) in Southeast Asia, with and without China
Source: Adapted from World Resource Institute 2003a.

Table 6.3. Southeast Asian Fisheries

FAO Designation	Pressure	Trajectory of Catch	Biodiversity	Other Signs about the Fisheries
About 15% overfished in the east Indian Ocean area	400,000 vessels in 1970 increased now to over 1.1 million—dramatic increase, much funded by ADB	Increased by over a million metric tons from 1992 to 2001; continued marginal increases expected	Indications of extreme biodiversity crisis—reefs near land-based pollution show 30–60% reduction in biodiversity; bombed reef areas show ~50% decrease in biodiversity indicating a dramatic and rapid decrease in fishery resources	Nearly all demersal fish near Philippines are overfished

Sources: FAO 2002; FAO 2003, Burke, Selig, and Spalding 2002; Bengwayan 2002; Edinger et al. 1998.

nomically overfished in almost all areas other than eastern Luzon, Palawan, and the southern Sulu Sea. (Burke, Selig, and Spalding 2002, 45)

This situation is directly related to destructive fishing techniques including the use of dynamite and cyanide to catch aquarium fish for the ornamental fish trade. Cyanide fishing began in the Philippines in 1957 on fish to be exported to the United States, and today the United States is still the largest importer of ornamental fish—85 percent of which come from the Philippines and Indonesia (Simpson 2001). This practice is confirmed in the Philippines and Indonesia, and suspected in Vietnam and Kiribati (Simpson 2001).

Catches in the Western Indian Ocean hit their all-time highs in 1999. The FAO (2002) believes that the Western Indian Ocean, Eastern Indian Ocean, and Western Central Pacific are the three areas where there is room for more exploitation since they have the lowest incidence of fully exploited, overexploited, or recovering stocks. At the same time, the FAO warns that these areas are the ones that are least understood and have the least reliable information, and that the "rapid increase" in deep-water fishing pressure is a significant threat to slow-growing and long-living fish which are easily disturbed. Also, 80 percent of the area's fish rely on endangered coral reef habitat, of which only around 10–30 percent remains intact. Some of the most richly biodiverse areas in the *world* will potentially lose 10–15 percent of their total marine fishery production for human consumption (Bengwayan 2002).

Coral Reefs

According to table 6.4, 82 percent of coral reefs in Southeast Asia face medium to high risk of mortality. This region holds 100,000 sq. km (34 percent) of the global reef area and therefore represents in absolute and relative terms the most square miles of reefs at risk compared to both the Caribbean and the South Pacific. According to the World Resource Institute, the United Nations Environment Programme-World Conservation Monitoring Center (UNEP-WCMC), ICLARM–The World Fish Center, and the International Coral Reef Action

Table 6.4. **Reefs at Risk in Southeast Asia**

Level of Risk	Low	Medium	High
Proportion of Reefs at Risk (%)	18	26	56
Global Reefs at Risk (%)	42	31	27

Source: Spalding, Ravilious, and Green 2001.

Network (ICRAN), 85 percent of the reefs in this region are threatened specifically by human activities (Burke, Selig, and Spalding 2002).

Sheppard (2003) notes that in the latitudes of 0° and 15° South these corals have a 20 percent chance of experiencing the same temperature as that in 1998, when 90 percent of the shallow coral died. Fifty percent and 85 percent of the Philippine and Indonesian coral reefs respectively (75 percent of the region's reefs are in these countries) are at risk at the same time staggering industrial and municipal waste is being loaded into their coastal zones (Jiang et al. 2001). In the last fifty years, the threat to Indonesia's reefs escalated from 10 percent to 50 percent, and it was at this same time (1960–1990) that the region's economy began to grow and become urbanized and globally connected. Indonesia's reefs, which account for half of the region's reefs, are most degraded in quantity and diversity in the western part of the country, exactly where most of the population and building occur. Twenty percent of these reefs are threatened specifically by coastal development, though most of the reefs also are specifically threatened by overfishing and destructive fishing practices (Burke, Selig, and Spalding 2002; Simpson 2001).

Climate-Related Change

Table 6.5 shows the climate changes found in Southeast Asia. Impacts of these changes are exemplified in events like the 1997–1998 El Niño which preceded the La Niña event in 1998–1999, in which Indonesia suffered very severe droughts which led to large forest fires. These forest fires created a large plume which advected over the Indian Ocean and was carried through oceanic currents (Paramaswaran, Nair, and Rajeev 2004). At the same time, the Southeast Asian coral reefs suffered a 90 percent mortality from the warming ocean waters, and these conditions are expected to continue to worsen (Sheppard 2003).

The monsoons in Southeast Asia are connected to ocean circulation, because monsoon climates occur where there is "strong seasonal contrast between land and ocean temperatures" (Bigg 2003, 78). The subcontinent of India and the

Table 6.5. Southeast Asian Climate-Related Ocean Changes

	Experienced	Projected
Sea Surface Temperature Change from the Mean	+1–2°C last 50 yrs; + 0.3°C per decade	+2–3°C with seasonal variations ranging to +6°C
Sea Level Rise	1.1–4 mm/yr	+5 cm next 50 yrs

Source: Sheppard 2003; Huang et al. 1997; Ali Khan et al. 2002; Banks and Bindoff 2003.

Tibetan plateau warm more than the ocean area in the summer months, creating a difference in air pressure, which ultimately brings moisture to the region. As the water temperatures change, we can expect that this phenomenon will also change, though probably in *un*expected ways given the complex intermixing of wind, heat, freshwater contributions, and other forces. Drought and flooding, the loss of mangroves, coral reef loss, human trauma through increased vector-borne diseases like malaria and dengue fever, and the likelihood that "several million" people could be relocated from the deltas in the region are expected (Harris 2003, 24).

Banks and Bindoff (2003) provide quantitative observations with modeled expectations of temperature-salinity changes in the Indo-Pacific. They note that in the Southeast Asia area between 10° and 24° North in the Indian Ocean, the water has become warmer and saltier over the last forty years, and that the surface warming of this water is the most important factor in changing the temperature (and therefore related qualities like density) in the lower water column. Areas above this latitude in the Northern Hemisphere are becoming colder and less salty as a result of freshwater inputs from surface sources in the northern latitudes. Banks and Bindoff test these results against historical perturbations and find that:

> all 30-yr changes starting from about 1960 exceed the 5% significance level from the internal variability. This suggests that the large-scale pattern of change seen in the observations of cooling and freshening on isopycnals in the subtropics of both hemispheres with warming (and salinification) occurring near the surface, may be thought of as a fingerprint of anthropogenic climate change in the ocean. (Banks and Bindoff, 162)

This translates as: Human-generated carbon emissions are pushing up the temperature in the Southeast Asia region, changing the way the ocean circulates in this area. The long-term results of this "tinkering" are unknown, but judging by the above synopsis of possible maladies, the future of climate change in Southeast Asia is very, very grim in human and ecological terms. Immediate protections for coastal people in this region, like housing located a safe distance from rising waters, storm surges, and mud slides, along with safe water sources, are needed to stem a massive loss of life from people who live in vulnerable, often informal, coastal housing.

Regional Political-Economic Conditions

Poverty and Violence

Table 6.6 shows that Southeast Asia boasts half of Asia's poor, and table 6.3 shows that over half of the region's people survive on less than $2 a day. At the

Table 6.6. Poverty and Government Expenditure in Southeast Asia

Country	% Population Living on Less than $1/day	% Population Living on Less than $2/day	Military Expenditure (% GDP) 2000	Health Care Expenditure (% GDP)	Education Expenditure (% GDP)
Region (average)	19	56	2.3	1.2	3.4
Bangladesh	29.1	77.8	1.3	1.7	—
Cambodia*	—	—	2.4	0.6	5.5
China	18	52.6	2.1	2.0	—
India	44	86.2	2.4	—	—
Indonesia*	7.7	55.3	1.1	0.8	1.4
Laos*	26.3	73.2	5.7†	1.4	2.4
Malaysia*	—	—	1.9	1.4	—
Myanmar*	—	—	1.7	0.2	—
Philippines*	—	—	1.2	1.5	3.2
Singapore*	—	—	4.8	1.2	—
Sri Lanka	6.6	45.4	4.5	1.4	—
Thailand*	>2	28.2	1.6	1.9	4.7
Vietnam*	—	—	2.7‡	.8	—

Source: World Resource Institute 2003a/b.
*ASEAN countries †1998 ‡1997

same time, military expenditures are, on average, over double the expenditures on health, and only marginally lower than education expenditures. The region leads in military expenditure and lags in health and education spending. Inflows of foreign direct investment in South and East Asia went from $12.4 billion in 1988–1990 to $63.2 billion in 1998–2000. This resulted from the liberalizing of the area's markets in the 1980s; this liberalization is now reversing to some degree in several of these countries (Bende-Nabende 2002).

In 1997, Southeast Asia was one of the central scenes of globalization gone awry, when the Thai baht currency was devalued as a result of bad loans funded from abroad for private ventures, mostly in failed real estate. The stabilization of the economy required international loans to the governments, which were paid along with guaranties of less government spending ("austerity" programs in the structural adjustments) on health care, education, and social policies. These cuts in services then were borne not only by individuals who had not profited from the initial real estate schemes, but by individuals who had had no say in taking the public loans from the IMF and its concomitant penalties for the poor when their social programs, little though they may have been, were cut to pay back and ensure the investments of wealthy Northern investors outside the region (Douglas 2002; Singh 1999).

The Asian Development Bank (ADB) is a primary force in these reforms. ADB's most recent primary focus is "poverty reduction," and the organization has an office dedicated to sustainable development. This is quite deceiving, however, since the ADB's policies appear to be, at least in some cases, enhancing poverty and aggravating environmental degradation in the Asian region, according to economist Raghav Narsalay (2001). The ADB is an important disseminator of global and regional foreign interests through policy and loans. For example, it is forced to discuss all of its policies with its partners: the World Bank, the IMF, and the Japanese government (Guttal 2001). Its approach to reducing poverty is to fund major infrastructure projects that allow for "trickle down" benefits such as projects that are "built-owned-and-transferred" (BOT) by private companies. These companies include giant corporations such as the United States–based AES Corp., which is the largest independent energy-producing company in the world, and many other giant transnational corporations described below.

Violence in the region is explosive compared to the other two regions. According to Eriksson, Wallensteen, and Sollenberg's (2003) data, since 1946, six of the region's countries have experienced fifty-six armed conflicts, compared to none in the South Pacific and one or two (if Colombia is included) in the Caribbean.

Fishing

Southeast Asia is the second-largest exporter of marine fish, surpassed only by the European Union. Southeast Asia has been supported by loans delivered every year since 1969 from the Asian Development Bank to increase capacity of the Southeast Asian fishing fleets and ports; this investment has led to the development of the largest fishing fleet in the world (Asian Development Bank 1995, 1997, 2002b; FAO 2002). As of 1997 the bank says it is changing its 1979 exploitation-focused policy into sustainable management of the region's "rapidly-declining fisheries," which are part of a "fish depletion crisis" (Asian Development Bank 1997, online). These are positive changes, but they do not address the fundamental issues of using an export-led infinite-growth model, or the issue of ecological distancing. They also come at least a full decade after it was recognized that the loans to Southeast Asian countries for commercial fleets were not sustainable for the fish and were hurting the small-scale fishers who made up the overwhelming bulk of employed fishers in the region (Bailey, Cycon, and Morris 1986).

The United Nations Food and Agricultural Organization (Vannuccini 2003) reports that China, Japan, Indonesia, India, Thailand, the Republic of Korea, the Philippines, Vietnam, and Malaysia are all among the top twenty world producers

Table 6.7. Armed Conflict in Southeast Asia, 1946–2000

Country	Minor Conflict	Intermediate Conflict	War
Burma/Myanmar	1960–1963	1950–1991; 1976–1988; 1993–1995; 1995; 1997–1999; 1997–2002; 2001–2002	1948–1949; 1964–1970; 1994
India	1967–1972; 1978–1988; 1989–1994; 1993; 1996–2002; 1995–2002; 1989; 1989–2002	1964; 1984; 1987; 1989–1990; 1992; 1994–1998; 1996–1998; 1992–2002; 2000–2002	1947–1948; 1965; 1971; 1990–1993; 1991; 1999; 1992–2002
Indonesia	1989	1991; 1999–2002	1990
Philippines	1970–1971; 1972–1980	1972–1977; 1979–1980; 1981; 1982–1988; 1987–1988; 1993–1994; 1994–1999; 1999–2002; 2001–2002	1978; 1981; 1982–1986; 1989–1992; 2000

Source: Eriksson, Wallensteen, and Sollenberg 2003.
Note: Each date represents a separate conflict; some dates overlap because there is more than one counted. Conflicts dated 2002 may still be ongoing.

in marine fish catch. It turns out that by the 1980s, Thailand, Indonesia, Malaysia, and the Philippines were leading recipients of fishing capital loans, and at the very same time export for total production tripled and "an increasingly high proportion of all fish landed in these countries was exported to Japan, the United States, and Western Europe" (Bailey, Cycon, and Morris, 1986, 1272).

In 1998, the ASEAN countries exported US$7.6 billion in fish products, and they continue to increase their pressure and efforts in the region. Officials from Oxfam and SEAfish for Justice (Mulekom et al. 2003), NGOs that work on social justice improvements and hunger eradication, note that at the same time fishing pressure and export rise, the region has sixty-five million undernourished people. As the pressure on fishing rises, subsistence fishers are having a harder time catching because more and more effort and technology (capital) are needed to get the same amount as before.

In fact, in 1994, the catch from five Southeast Asian countries—Indonesia, Malaysia, the Philippines, Thailand, and Vietnam—represented 97 percent of the total catch in the Western Central Pacific (which encompasses mostly what I am calling the South Pacific region) (FAO 1997). This area is eclectic in its landings. Fisheries are exploited so that the top catch is "miscellaneous," followed by tunas, jacks, herrings, redfish, mackerels, and shrimp (FAO 1997). Increases in capital concentrating on increased fishing ability have resulted in a dramatic increase in trawlers starting in 1970. At the same time, the rate of increase in fish catch flattened out in 1990 after increasing since 1950 (FAO 1997), and this is correlated with the loss of demersal fisheries in the Gulf of Thailand, which now stands at one-tenth (a 90 percent loss) of what it was when the capital increases began in that area (FAO 1997).

Japan is the foremost fish importer in the world (about 23 percent), with the United States (about 17 percent) coming in second. These two countries also lead the world in fishing subsidies, at $2,935,300 and $867,900 respectively as of 1997 (Yale Center for Environmental Law and Policy 2002). Thus, almost half of all the fish caught in the ocean go to one of two countries, with another 30 percent going to other industrialized nations. Fifty percent of the world's traded fish come from Low Income Food Deficit Countries (LIFDC), which provide almost $18 billion in net revenues (subtracting import values) and outweigh all other agricultural commodity exports from poorer countries (Vannuccini 2003).

While China is, by far, the largest producer (catcher) of marine fish, Thailand is the largest exporter of fish in the world. A great deal of Thailand's exports are shrimp-related commodities (Vannuccini 2003). Shrimp farming is also responsible for clearing vast areas of mangrove, which is a serious and growing problem in the region. In total, shrimp represent just under a fifth of the world's fish export value (Vannuccini 2003). Hundreds of thousands of hectares of mangroves have been developed with global and regional finance (see below), specifically via the Thai firm Charoen Pokphand (CP Group), which now dominates the development of shrimp farms in Indonesia. Juan Martinez-Alier (2000) argues that this is consumption-driven globalization without large TNCs.

However, while shrimp production is clearly consumption driven, some of the world's largest corporations are in fact involved or have been involved in developing intensive shrimp farming. Also, the industry is "characterized by a high degree of vertical integration, usually organized by the companies which supply feeds and laboratory services associated growers" (Goss, Burch, and Rickson 2000, 517–18). For example, the CP Group has worked with Mitsubishi to develop the shrimp-farming technology, and the International Basic Economy Corporations (USA) and Kentucky Fried Chicken (USA) have both been pivotal in the CP Group's vertical growth. Further, the CP Group's joint venture with Continental Grain (USA) to establish feed mills in China was fostered specifically through the recent global economic expansion in shrimping after the industry collapsed in Taiwan in the 1980s. Thailand received US$84 million in assistance for shrimp development and a US$11.1 million loan from the ADB in 1986, and the World Bank and the Bank of Thailand provided seed funds for these projects (Goss, Burch, and Rickson 2000). The CP Group is the world's leading shrimp exporter.

It is worth noting that there are some efforts to revert to the decentralized traditional fishery governance found in most Southeast Asian countries prior to colonization. Indonesia, for example, centralized its fisheries from 1966 to 1998, when most growth of export and capital occurred. Now, under the 1999 Local Autonomy Law, locals have more responsibility, and some traditional community-based management is being recognized with some problems, but by and large, decentralizing the fisheries is a much preferred and more efficient and equitable approach (Satria and Matsuda 2004). This comes at the same time that the country as a whole struggles with democratic reforms and its authoritarian tendencies.

Coastal Development

Coastal development is one of the key threats to sustainability in this region, which has dramatically changed in the last fifty years. Huge amounts of coastal wetlands have been eliminated: Cambodia (100 percent), Indonesia (46 percent), Malaysia (59 percent), and Vietnam (98 percent) (Burke et al. 2001). In the last fifty years, these same countries have also suffered as high as 75 percent losses of their estimated original mangrove cover (Burke et al. 2001). The most important threats to nonsustainable coastal development are the intensification of shrimp farming, noted above, and rapid urbanization. Both trends are discussed in detail here.

The meaning and impact of this coastal development were perhaps never more clear than on December 26, 2004, when a devastating tsunami hit the region, in particular the Aceh region of Indonesia. Over 200,000 people died in the Indian Ocean region. However, in this tsunami and in a "supercyclone" that struck India in 1999 and claimed 10,000 lives, fewer people died where mangroves and coral reefs were intact and healthy because these "bioshields" can absorb "the fury of the waves" and the energy of a storm (Padma 2004, online). But these areas are among the most threatened in the world for reasons detailed below, even though they save lives and provide basic ecological goods and services essential to coastal people.

The shrimp farms rely on the conversion of mangrove forests in the coastal zone where the shrimp ponds are built. In Thailand alone, shrimp pond areas have increased from 9,056 ha in 1972 to 91,887 ha in 1993 (Goss, Burch, and

Rickson 2000). Martinez-Alier (2000) says that governments in Sri Lanka, Thailand, Indonesia, India, Bangladesh, the Philippines, and Malaysia (in addition to Ecuador and Honduras) allowed and continue to allow for the enclosure of mangrove commons in order to capture a piece of the $10 billion world shrimp trade, which goes almost entirely to private firms and individuals.

> The mangroves are usually public land in all countries, being in the tidal zone, but governments give private concessions for shrimp farming or the land is enclosed illegally by shrimp growers. Illegality is prevalent not only because of the public character of the land, but also because there are often specific environmental laws and court decisions protecting the mangroves as valuable ecosystems. (Martinez-Alier 2000, 1)

This illegal process was supported directly by the World Bank until the mid-1990s (Martinez-Alier 2000), and in some cases was fully directed by the World Bank—almost entirely for export of the shrimp to rich consumers—as part of loan conditions despite inhumane working conditions, land evictions, and ecological disasters in places like Bangladesh (Ahmad 1995). Martinez-Alier argues this process is replacing a typically sustainable structure of livelihoods that are dependent on the mangroves and their ecological services, which include fish and shellfish nurseries and sources of building materials and charcoal which are directly used and sold in small-scale economies by locals. Loss of mangroves reduces protections against sea level rise, and the pollution from these shrimp farms is seriously damaging fisheries in the area, as well as contributing to the loss of nursery space (Martinez-Alier 2000).

While the shrimp ponds are often owned by small farmers, 60 percent of the shrimp feed is controlled by a few corporations—namely Cargill of the United States, Charoen Pokphand of Thailand, and Japfa Comfeed of Indonesia. Cargill is also a "knowledge systems" consultant to the World Bank (see www.worldbank.org/ks/KSclient.html), and Cargill Fertilizer has had a $7.3 million contract for agriculture and natural-resource projects (Asian Development Bank 2002a). Phosphate Chemicals Export Association (PhosChem)—the most important vehicle for U.S. export of phosphate chemicals—has had at least $4.6 million in ADB contracts, and includes as a member the Potash Corporation of Canada, which is the world's largest fertilizer producer, has the world's largest capacity in phosphate feed ingredients, and is the world's largest producer of industrial nitrogen products (Potash Corporation 2004). All of these industries have a vested stake in the growth of shrimp farming, and many have had direct involvement in pushing the industry forward despite the socioecological wreckage it creates.

One of the most important corporations in shrimp farming is the CP Group, which is the largest agribusiness conglomerate in Asia with interests in petrochemicals and telecom, hog, chicken, and shrimp farming, and a host of other interests all around the region (Lee 2003). It is the world's largest producer of animal food and tiger prawn shrimp (*Economist* 2001). The World Rainforest Movement (Lang 2001), an international NGO, reports that Charoen Pokphand has worked with the ADB and the World Bank to expand the shrimp export throughout the region. The World Rainforest Movement links this expansion to the loss of mangroves, specifically in Vietnam where tens of thousands of acres of rice paddies have been converted into shrimp ponds under heavy domestic and international loans. Ironically, the organization reveals that the World Bank, at the same time it is producing pressures on mangroves, is lending Vietnam $31 million in a project to rehabilitate them. In other words, public funds will be used to repay the loan which was needed because of these structural decisions to promote the illegal private shrimp farms for export to wealthy consumers. A similar loan was made to Indonesia for the rehabilitation of coral reefs in North and West Sumatra (Asian Development Bank 2004). Since Japan and the United States benefit the most from lower shrimp prices, it is not surprising that they are also the largest investors in ADB. Asian shrimp make up 90 percent of the U.S. market, where shrimp wholesale prices have dropped 40 percent from 1997 to 2002. This comes directly from the difference in modes of production in coastal shrimp farming from coastal trawling (Fritsch 2004). Thus, the ecological prices, which are higher, are shifted to the coastal poor, while the economic prices for the shrimp actually go down as a result of neoliberal focus on large-scale exports.

Down to Earth, a United Kingdom–based ecological justice movement for Indonesia, reports that several companies—PT Central Pertiwi Bratasena (31 percent owned by the CP Group), PT Dipasena Citra Darmaja, and PT Wahyuni Mandira Company—are using a controversial method of shrimp farming called the "nucleus estate" small holders scheme (NESS). This is a process whereby a large company converts the land into shrimp ponds and then creates agreements with small holders to operate the farms. The result is supposed to be a payoff around seven or eight years later, which is dubious given the five- to ten-year life span of a pond, by the small holder. However, World Rainforest Movement and Down to Earth both argue that this has destroyed hundreds of thousands of hectares of mangroves, put many people in deep debt, allowed for illegal takings, and spurred on riots, murder, and violence against women.

Another concern is urbanization in the coastal zone. A thorough analysis by Jiang et al. (2001) of coastal development throughout the region informs this section. They note that in 1950, there were only two "megacities," or cities with over 8 million residents. As of 1995, there were twenty-three, twelve of which were in Asia, and most of these are in South Asia. This development is tied to the expansive population growth that has occurred in the region: in 1980 there were over 700 million urban residents in Asia, and now there are over 2.2 billion. As many as 420 million of these residents are not connected to sewage. Consequently, human waste flows through open ditches to the nearest river and coastal area where most cities are located. In Jakarta, there are over 30,000 factories that discharge waste into the rivers that flow into the bay at the same time that zero percent of the residents have a sewer connection (Jiang et al. 2001), and "[a]n analysis of seven countries in the East Asian Sea (EAS) shows that most wastes enter the sea directly or through rivers, canals, and drains without any treatment" (Jiang et al. 2001, 306).

Also, port access for global container ships has had significant impacts. As Singapore builds and keeps up its global port, its reclamation of coastal areas has affected 60 percent of the country's reefs, while increased sediment loads have decreased the amount of coral in all monitored sites. Reefs that are doing well are connected to islands and microareas that are isolated and not connected to development, urbanization, or heavy trade: the Makassar Straits, Flores Sea, Banda Sea, and areas off the Andaman Islands, West Papua, Myanmar, and Thailand face far less stress (Burke, Selig, and Spalding 2002).

Thus, in the Southeast Asia area, coastal development is occurring on a very large scale, with massive oceanic impacts. Jakarta, Manila, and Bangkok all generate large amounts of industrial and hazardous wastes, and all of these cities are on the coast. Manila Bay boasts "smoky mountain," which is a reclaimed 34 ha area of urban waste that rises 40 m above sea level and smokes with carbon monoxide and methane gases (Jiang et al. 2001). These areas are experiencing increased sediment loads from the rivers, which bring high levels of heavy metals such as lead, cadmium, and mercury, which are all toxic. This material is spilling over into fisheries and creating important economic and health costs. For example, fish and shrimp from Jakarta Bay have been found to have 3.7 micrograms of mercury per gram of meat—700 percent more than the allowable limit as set by the World Health Organization (Douglas 2002). Also, these urbanized areas in Southeast Asia are seeing an increased built environment on or near wetland areas and beaches where severe loss of mangrove, seagrass meadows, and coral reef area is a direct result (Jiang et al. 2001). In fact, Jiang and others note that if continued rates of coastal development and pollution continue in the region, the entire habitat of mangroves in this region will be lost by 2030. Thus, the current trend of urban coastal development, as it is occurring now, is one of the more unsustainable features of the region, and is the most severe case of coastal-zone decline treated in this book.

Poor urban environmental conditions have direct connections to global neoliberal political economic sources because international agencies are focusing on economic growth instead of addressing monumental urban hazards in addition to local trenchant opposition to progressive policies. Hardoy, Mitlin, and Satterthwaite (2001) point out that in urban areas of the global South in general, access to water and sewer systems, marginal and informal housing for the very poor, biological agents in drinking water, severe hazardous pollutants produced by mostly unregulated industries, and the causes of vector-borne diseases are relatively ignored by international funding. Only 2 percent of ADB credits go to the poor, for example (Guttal 2001).

> It should be noted though that the ADB's approach to making the market work for the poor is not through support for domestic regulation and protection, or through ensuring adequate opportunities, access and services for the poor, but rather, by expanding the private sector's share in physical, financial and social infrastructure. (Guttal 2001, 5)

This funding is going to coastal development projects that are damaging fisheries and increasing pollution and sediment in these areas—all of which are important factors in coral reef degradation. For example, the US$750 million Samut Prakarn Wastewater Treatment Facility in Thailand is expected to degrade social and environmental conditions severely:

> The site is expected to produce 50 tons of sludge everyday, which would flow into the sea and destroy local fisheries and mussel farming, resulting in a serious decline in living standards of people in and around the area. Equally serious are charges that the land for the wastewater plant was acquired by less than legal means. (Guttal 2001, 7)

Another example is the Theun-Hinboun dam project funded by the ADB in central Laos, which has affected between 4,000 and 5,000 people (some estimates go as high as 10,000 people), who have suffered due to the loss of fisheries, clean water, and other key subsistence sources (Adams, 2001).

Knowledge Production

Poverty is one of the main barriers to knowledge production in Southeast Asia. Poverty alleviation programs, such as through the ADB as noted above, are apparently worsening the problem. Even when money is generated specifically for (as opposed to trickle-down processes) reduction in poverty through direct social investing, poverty remains desperate. For example, Indonesia's absolute poverty did not decrease in the late 1990s, despite ADB loans for social programs (Hadad 2001): "The UNICEF Jakarta office has stated that 'due to its serious debt burden Indonesia would sustain a lost generation, a weak and feeble-minded generation resulting from malnutrition, lack of education, and unhealthiness'" (quoted in Hadad 2001, 46).

Much of the social-investment money went to 1998 election campaign funding and to the militia fighting in East Timor at the time. Further investment has apparently not helped either, and one wonders if this is because these efforts are countered by larger public loans that require austerity programs, privatization, and the loss of government social safety nets (Hadad 2001).

Technological development is a key component of knowledge production, and almost all of this knowledge is concentrated in OECD countries. Ninety percent of all technological and product patents are held by transnational corporations which are headquartered in ten OECD countries, so that most technology transfer is done through mergers and acquisitions (Douglas 2002).

Another main source of knowledge production comes from academic contributors. One source of knowledge about Southeast Asian sustainability comes from the Southeast Asia in Transition project (www.seatrans.net), which has regional partners but is operated, directed, and funded from Europe within the Institute for Interdisciplinary Studies of Austrian Universities. This project provides indicators of "material flow accounting" (MFA) and other social-environmental knowledge relating to the metabolism of Thailand, the Philippines, Vietnam, and Laos. This group provides pivotal information about the region, including some methods, like the MFA, for accounting for sustainability, as well as reports, such as the Martinez-Alier (2000) essay noted above, which is publicly available.

Some local knowledge building about sustainability, however, comes from Southeast Asian voices, even if these voices are in the minority. Much of the knowledge building is coming from academic researchers who ally with the growing number of environmental NGOs (especially at the country level) (Subhanrao Pednekar 1995), and can be seen in cases like the ASEAN Regional Center for Biodiversity, noted below as a part of the regional civil society sector.

Other knowledge production about ecology in the region is directly related to industry interests in marine exploitation. For example, the Southeast Asian Fisheries Development Center Aquaculture Department conducts research related to the profitable food fish market. This research includes larvae and fry studies on commercially important species in the area. Among these species are endangered ones such as the seahorse and the tropical abalone (Marte 2003).

Regional Institutions and Social Affiliations Affecting Global Politics

Since the era ranging from 1960 to 1990, the region experienced heavy economic growth, and absolute poverty fell in the countries that experienced this growth (Warr 2001). While this is extremely important, this is a very modest change. The overall percentage of poverty went from just under 50 percent of the region to around *40 percent of the region, and on average, the poor person had $1.08 to use per day to survive* (UNESCAP 2003). During the late 1990s, these very same countries were in recession and poverty grew again in some of these countries (UNESCAP 2003). This poverty fluctuates with the foreign investment which has driven the region's economic growth. Thus, growth helps to reduce poverty in this area. At the same time, creating neoliberal economies demonstrably increased the vulnerability of the poor in these countries who apparently are not protected from the volatilities of this growth. When these economies required aid orchestrated by the region's elite in each country, "their impact was more widespread and felt most acutely by the poor" (Beeson 2004, 448). And, at the best of times, extreme poverty still affected 40 percent of the population.

The most important regional institution in Southeast Asia is clearly the Association of Southeast Asian Nations (ASEAN), which was originally founded, like many regional bodies, for security reasons and originated under the auspices of British military planning during World War II (Beeson 2003). However, ASEAN is now almost entirely focused on economic cooperation, and from the beginning grew out of politics "facilitated by the common shift of its members towards authoritarianism and reflected non-democratic values" (Acharya 2003, 388).

ASEAN was successful in turning an area that was previously known for war and strife in the postcolonial era into a more peaceful and stable region, and thus it was successful in creating a genuine regional identity (McMahon 1999; Narine 2002). However, the region will probably not be able to transcend the current desire of noninterference between states in the region despite a trend toward more democratic norms generally (Narine 2002; Acharya 2003). This means that the region will not be as cohesive as, for example, the Southern African Development Community (Thompson 2000) or the European Union for some time because these regions were able to ameliorate some of the norms for absolute sovereignty (e.g., absolute noninterference) for the greater regional good. ASEAN, therefore, will likely be marked by internal tensions, including the powerful influence of the

United States, which will mean it will consistently be reacting to external influences rather than creating its own autonomous agenda (Beeson 2003).

In spite of this situation, there is an offshoot of ASEAN—the ASEAN Regional Center for Biodiversity (www.arcbc.org)—that has begun to play an important role in coastal sustainability. This is a regional organization made up of international representatives from ASEAN countries, and is funded through the European Union. The organization maintains what it calls the "most comprehensive" online database for biodiversity for Southeast Asian countries and provides important assessments of environmental concerns in the area. It has done important work on integrating coastal management with cities, and with marine protected areas (MPAs) which have various restrictions, such as on fishing or development, and are an essential conservation tool in such a sensitive area with so many pressures. It acknowledges that the region's focus on a cash economy, urbanization, and intense population growth is the underlying cause of unsustainable coastal development and reef threat. The center counted 310 MPAs in 9 regional countries, but says that 80–90 percent of the MPA are not managed effectively and that about half are not managed at all (ARCBC 2002). Still, as the region encroaches on the coastal zone, MPAs will become increasingly important for mitigating impact, and it is a good sign that they have increased in number recently.

The role of civil society in the region shows mixed trends. In some countries like Cambodia, environmental NGOs struggle for permission to organize (Subhanrao Pednekar 1995). Further, Cambodia, Indonesia, and Burma all have a history of using "disappearances," or politically motivated murders typically conducted by the state, torture, imprisonment without trial, and executions without trial (Sponsel 2000). Perhaps worse, there is inadequate knowledge about these issues in Brunei, Singapore, Thailand, and Vietnam (Sponsel 2000). J. S. Eades (1999) believes that this trend in authoritarian and paternalistic centralized government runs concurrent with economic priorities of the state. Eades believes that when high-speed economic development in the Asian-Pacific is in its beginning stages, highly coercive government regimes are formed to ensure the continuation of these policies. As such, environmental movements are dormant, repressed, or nascent at best.

In this region,

> welfare policy has been dominated by economic rather than social considerations supported by some underlying ideas of anti-welfarism and, especially, by resistance to the provision of government-guaranteed social welfare. The ruling elites have generally only accepted the institutional

concept of social welfare when confronting a political crisis; when this is overcome they return to the "residual concept of social welfare." (Park quoted in Dragsbaek Schmidt 2000, 223–24)

This is due not only to Western influences, but also to a Confucian ideology where social control, stability, history, genealogy, and authority all have a strong influence.

Repression of citizens occurs while governance is relaxed for corporations. Indigenous peoples, such as those relocated from Java to Sumatra, and the so-called sea-gypsies or boat-people (those who live permanently on makeshift boats) are particularly threatened, violated, and politically marginalized wholesale in the region. This repression of the region's civil society is linked to ecological degradation (Sponsel 2000). Unfortunately, neither the West, ASEAN, or other major groups have offered effective protection against severe repression to suffering citizens such as boat people, Timorese, Burmese, or Chinese (Knippers Black 1999).

At the same time, strong pressures to democratize in Southeast Asia are coming from South Korea, Taiwan, and other economic Asian leaders. There have been important democratic movements, for example, the "people's power" movement in Manila, which was mirrored in similar demonstrations in Bangkok and Jakarta (Hedman 2001). Also, as the region has urbanized, civil groups are "crystallizing into non-governmental organizations" (Douglas 2002) where environmental/sustainable development groups with *regional* social justice agendas, such as the Asian Coalition for Housing, have been formed in nearly all of the respective countries (Douglas 2002). In this way, Douglas (2002) argues that there is a serious push for livable, more democratic, and sustainable cities in the region. Most of the civil-society environmental groups appear to be working at the local level—for example, artisanal resistance to government-sanctioned mangrove clearing, or the areas which have declared themselves "shrimp free zones"(Martinez-Alier 2000; Subhanrao Pednekar 1995).

Intellectual Walden Bello sees a worldwide antiglobalization movement in the region "fusing" with antiwar movements (quoted in Beeson 2004), but Beeson believes these movements are heavily mitigated by authoritarian state impulses, and are mostly reactionary, ad hoc, embryonic, and having limited effectiveness. Nonetheless, Beeson makes the important point that increased unilateral militarism from the United States' "war on terror" has deepened elite and popular resentment, embedding its (the United States') own contradiction and ultimately limiting the power of the United States and neoliberalism in the region in the long run.

Comparison to the Borgese Test of Sustainability

Clearly, Southeast Asia is not on a sustainable path. This chapter shows that there are specific combinations of ecological sociopolitical problems that are aggravating one another. Ecologically, the only indicator that is not pointing universally downward is the region's fisheries, and many of these fisheries are in desperate trouble—such as those of reef, demersal, and coastal-dwelling fish—even if the region on the whole is not overfished by FAO standards, which I am forced to see as counterintuitive.

The rest of the indicators are grim. The region is the ocean's most biologically diverse at the same time that it is the ocean's most threatened. Coral reefs are severely threatened, the coastal zone is much damaged, and both sea surface temperature and sea level are rising. These changes are expected to cost the region billions of dollars in ecosystem goods and services, but these costs will not be borne equitably. Food, resources, storm protection, sea level rise protection, and other aids that will decline and are declining will affect mostly the poor. Bangladesh lowland coastal areas are expected to require the relocation of thousands of people, and nearly 80 percent of that country lives on less than $2 a day; this is a combination of forces that is not encouraging. Immediate action is required to stem massive loss of life.

At the same time that these losses and burdens will be placed more heavily on the poor, just fewer than one in five people in Southeast Asia are under the most severe international poverty line and live on less than $1 a day; this level is clearly unsustainable, even if this number has decreased. Also, in this region there are fifty times the amount of armed conflicts than in either the Caribbean or the South Pacific. This latter condition could be driven by the higher levels of arms expenditure, though traditional international-relations scholars may see it the other way—that the danger of the region is driving the expenditures. Either way, it is clear that the region that spends the most on arms is also the most war torn and the most ecologically damaged. I do not see these conditions as accidents.

The ability of Southeast Asians to respond to these problems is severely compromised by the neoliberal globalizing forces combined with parallel Confucian-influenced local interests, which seem to be aggravating many of these problems, if not outright causing them. National governance is tightening and becoming more authoritarian, there have been frequent armed conflicts, and social hierarchies are very steep. Consequently, it is a dangerous business for the growing civic groups to reproach authority. Neoliberal banking institutions insist on austerity programs that leave the region spending an average of 1.1 percent of its GDP on health care, though a stronger 3.4 percent for education tops the 2.4 percent military expenditure. Loans to the region are burgeoning, and are paying companies for increasing ecological pressures. Thus, Southeast Asia's poor will have fewer social provisions while their governments become indebted to international finance for bankrolling some of the world's largest multinational corporations, such as AES and LaFarge (cement), for programs that are geared toward more industrialization and growth in exports. Such a focus apparently has not done much for raising the conditions of the worst off. As these ecologies are changing, as is seen in the case of corporate welfare for shrimp farming, the poorest lose land and ecological services, and gender inequity and suffering increase (Siregar 2004).

This is the widest sustainability gap for any of the regions discussed in this book, and it leaves quite a dramatic degree of public policies that would be helpful to reposition the region on a more sustainable path. As Michael Bengwayan (2002) of the *Asian Observer* poignantly writes of the Philippines, the region is "fast becoming the land of the dodo" in ecological and social terms. He points out that the top four countries with the highest number of threatened species on the IUCN Red List (the list for globally endangered species used by the United Nations and most scientists) are *all* in Southeast Asia. So, governments should first abandon public loans that saddle them with requirements to decrease services to the poor. Aid and loans that are accepted should be aimed directly at social-service programs that provide a consistent, reliable, and meaningful safety net, such as minimum income benefits, so that no one is left with less than subsistence living, especially women who may not be able to work for the same wages as men. Immediate protections for housing and basic provisions should be deployed for low-lying coastal people, *many of whom will die* if this is not done. Storms will come, along with surges, flooding, and mudslides, and there are fewer natural barriers from the reefs, roots (of trees for stable soil), and mangroves that might have otherwise protected some people. If this is not a sign that sustainability has been overshot, I do not know what is.

Second, comanagement policies should be adopted to reallocate the coastal areas as commons with genuine protections. Marine preserve management should receive funding from European nations and perhaps tourism fees; the more of these MPAs that can be created and well managed, the more fecund surrounding areas will be. Other options for protecting against damaging coastal development should be explored, such as sustainable-scale use areas like the "extractive preserves" in Brazil where building is severely limited or halted on an industrial scale, and extraction is limited and governed by the locals who are most affected by declines.

Synthesizing Ecological and Political Structures to Their Global Structure

In sum, the depletion of fisheries has very strong links to international loans made to increase commercial export mostly for the United States, European Union, and Japan. Local, regional, and global coastal development, neatly embodied in the example of shrimp farming and including intense urbanization, is damaging the coral reefs. All of this is occurring at the same time that global foreign investment increased more in this region than any other studied here.

Southeast Asian water temperature and current are changing, and this threatens to change regional weather patterns. The region is contributing substantially to climate change, but it simply does not have the historic contribution of greenhouse gas emissions of the West, and especially the United States, because industrialization came later to the region as a whole. China in particular threatens to increase its absolute and per capita emissions rates, but in both terms China lags substantially behind the United States. The assumption that the World Ocean will distribute global warming factors globally is confirmed; there was little reason to have doubted this in the first place, but I make the point here because it also reinforces the theoretical position that there is sufficient reason to see the world water body as a holistic World Ocean.

In conclusion, there are several correlating structural factors in this world system from a political-economic and ecological perspective. Change in the region-level FDI correlates regionally with coastal development, the ecological decline of reefs and fisheries, and biodiversity loss. If we had eliminated the regional perspective and looked only at countries individually, the commercial hubs in the region would have been hidden. From a regional perspective the flows of global capital and their impacts on surrounding interconnected ecology are much easier to see. Violence also rises with a change in intensity in globalization. So does a loss of civic local and regional power since each region's state and regional regimes weakens as the rate of globalization increases based on the comparison among three regions.

The direction of trends for social and ecological health in Southeast Asia does not inspire confidence. It is possible that interregional political resistance will be able to replicate the success found in the South Pacific, but right now social and ecological considerations in Southeast Asia are blotted out by economistic neoliberal policies enforced, very often, through the draconian power of the State while international business is given a free pass. This trend is behind a globally connected exploitation of people on the lowest rung of political power and an overshoot of the region's resources which themselves have global import in a World Ocean system.

Connecting the Parts—Theoretical Connections 7

IN THIS CHAPTER, I take the findings from the three regional cases and combine them into a global picture of ocean sustainability using the three theoretical perspectives described in chapter 2: complex systems theory (CST), hermeneutics, and critical theory. I conclude by identifying common threads among the different perspectives to see if there is any consilience among them that would indicate an intersubjective consistency. In tables 7.1 and 7.2, the regional ecological and social trends are summarized.

Complex Systems Theory (CST)

CST has particular advantages for interpreting sustainability for the World Ocean. By CST, I mean a theory of evolution that describes how elements—which are often simple—combine and change through networks of interdependent reactions and self-organize in nonlinear bursts (Capra 2002). Nonlinear systems evolve toward critical states, which are seen as spontaneous intermittent fluctuations; order is observed at all scales derived from small perturbations (Seuront and Spilmont 2002). In this section I describe a theory of the global through CST and the sense of sustainability we are able to garner from this insight. First, I review some of the findings in this book in the context of CST by describing their apparent complex characteristics, emergent properties, and the structure of the ocean ecosocial systems.

The World Ocean is complex because simple elements (e.g., simple chemical elements of carbon, oxygen, hydrogen, and nitrogen) evolve into fantastic combinations in nonlinear relations originating from small or local perturbations that create global patterns. At each of the regional locales, we see specific perturbations which have coherence at the global level. There is a global pattern of simplification through the elimination of reefs and fish, and the range of fish that are allowed to exist in robust populations. Changes in water temperature, movement, and salinity are occurring in very small degrees; but these will gain and have

Table 7.1. Ecological Summary

Region	FAO Fisheries Assessment	Fish Biodiversity	Other Indications of Fishery Health	Fishing Capital	Coral Reef	Coastal Development (low, moderate, high, or intense)	Sea Level Rise	Sea Surface Temperature Change
South Pacific	8% overfished	Signs of biodiversity problems not clear	Serious decline not yet evident beyond a few cases	Moderate, global capital; increasing regional capital	41% med/high risk	Low intensity, local and foreign capital. Mostly intra-regional capital through Aus.	0.85 mm/yr; +5 cm next 50 yrs	+0.12°C on average, with some areas actually cooling; the warmest section at 24°N with +0.1701°C
Caribbean	30% overfished	Serious threats evident	Coastal and reef fisheries overfished	Low degree of capital, locally based effort	61% med/high risk	Moderate foreign capital through tourism	20cm/100yr; +6 cm next 20 years	+1°C last 50 yrs
Southeast Asia	15% overfished Reasons to believe this undercounted	Cataclysmic threats evident	Coastal and reef fisheries overfished	Intense, global capital	82% med/high risk	Highly intense, global capital	1.1–4 mm/yr; +5 cm next 50 years	+1–2°C last 50 yrs; +0.3°C per decade

Table 7.2. Regional Political Summary

Region	Are the formal international institutions seriously planning for social, political, and ecological sustainability?	Knowledge production locally generated and beneficial, and what is the impact of indigenous knowledge systems?	Is the degree of poverty in the region severe?	Armed Conflicts 1946–2002	Interpersonal institutional hierarchy (steep, moderate, or egalitarian)
South Pacific	Social, political, and ecological	Mixed locality, highly interdisciplinary, notable impact from indigenous knowledge in SPC and FFA	Probably	0	Tends toward egalitarian, moving toward a moderate model
Caribbean	Ecological and economic concerns	Mixed locality, not notably impacted from indigenous knowledge in CARICOM	Yes	2 (United States)	Moderate to steep
Southeast Asia	Economic sustainability only	Mixed locality, not notably impacted from indigenous knowledge in ASEAN	Yes	56	Steep, several authoritarian regimes

gained fantastic and seemingly irreversible momentum, such as in the thermal expansion of ocean water and changing weather patterns. All three regions show differing degrees of change, but all three ecological conditions show a pattern of decline. This will inform the evolution of a new structure in the holistic World Ocean. There are several things we should expect even if prediction is difficult or impossible.

Since the oceans are interrelated with human subsistence and global trade, especially in fisheries and carbon emissions, human communities, particularly those in coastal areas, should expect that there will be abnormal conditions in the future that will change in ways not yet knowable in the material and energy through fisheries, water properties, and temperature change and circulation. Small changes can set off large ones; and in a global ocean, many of these changes will be well mixed, flowing through the open ocean ecologies to create global patterns of change in climate and the organization of life.

The number and quality of ocean nodes—ecologies including human uses and interactions—are important factors. As part of the ocean system, biological organisms consist of both matter and energy that connect through nutrients, food, chemicals, and heat to other biological organisms and other matter and energy.

These ecologies demonstrate several trends. One trend is decreasing biodiversity in the ocean in fisheries and loss of coral reefs around the world, and therefore a simplification of the matter and energy that organize the larger ocean subsystem and Earth system. Climate-related nodes in the ocean are changing. Heat energy stored in the ocean is increasing as a matter of global warming. Carbon is increasing in volume in the atmosphere and ocean water and the gas exchange for algae has likely reached its maximum. Material (water, salinity) and energy (temperature) are changing worldwide. Since oxygen and carbon are relatively well mixed at the global level, and the ocean is transferring its heat around the world, these changes are universal within the World Ocean.

Sea level rise is occurring throughout all regions, along with ocean temperatures, though the impacts vary; for example, polar warming is increasing the cold-water freshening in the Southern Ocean. Rising ocean temperatures carry a "momentum" that will create well into the future fast, deep, and lasting climate changes that will probably be nonlinear and in some cases unexpected. We know that ocean-related climate changes are occurring and these are very likely a result of very fast human climate forcing, but we do not know what this will mean generally for fisheries, storms, drought, fires, and other troubling issues even fifty years from now—a mere blink in geological time. Positive feedback elements, or those that reproduce and enlarge the process of change, add to the "patchiness," or intermittency, of resources, and increase the probability of "catastrophic" results (Rietkerk et al. 2004). We do seem to know that coral reefs will likely suffer extensive mortality worldwide from these climate changes, and we know that this will have a devastating impact on fishery production. However, we do not know the resilience of fisheries and coral in general to these changes or their ability to generate resilience, or what these severe biological losses will mean for the rest of the material and energy nodes that interact with the coral.

Socially, complexity offers potential hope that the pressures exerted in a system may create unique responses. Civil society and counterhegemonic activity, such as creating class-conscious ecological movements, seem to be springing up in nonlinear ways, but are still not comparable to the scope of power found by firms or states within the world capitalist system. There is not a direct relationship that is predictable to the way civil society organizes around these ocean-related changes, but there is a global civil society struggling to organize, and the ocean provides a context for this organizing. Internet connections allow for knowledge production and public relations to be mobile, even if labor is not.

In fact, knowledge production indicates one of the biggest hopes for sustainability in all of the regions. In every region, there was evidence that knowledge production was tied to and funded by, and in some cases partially controlled by, former colonial powers and current centers of capital in the United States, Europe, or Japan. Yet, despite pressures, every region also had pockets of local knowledge-building programs based on traditional, "old coastal," tribal, or other small-scale fishers and focusing on marine natural resources, ocean sustainability, and to a lesser extent, comanagement schemes. The way this knowledge is transferring to modern institutions appears related to the adoption of neoliberal reforms, where the South Pacific—least touched by these reforms—is transferring traditional knowledge more readily, even if not easily or without conflict. Thus, despite countervailing forces, people everywhere are organizing their own inquiries and deriving their own knowledge about the ocean systems that surround them. This gives me hope, particularly from a view of complex systems theory, because this knowledge building will inform and disturb the current economic trends. Complex systems theory reminds us that big changes can come from small bits of information under conditions of connectivity.

This gets to probably one of the most critical questions facing regional NGOs—whom will they represent? Clearly this will vary, but simple environmental consciousness is not enough to justify a "monocropping" of one type of NGO, such as the Northern conservationist-oriented NGO. NGOs evolve over time, and conservation is becoming more locally sensitive in pragmatic terms. But, there is an important opportunity for regionally based NGOs to specifically represent nonaffluent local interests that are more often "being globalized" than they are "globalizing" (Dobson 2003). On the other hand, there will be monumental temptations to represent the more affluent and powerful interests from the North in ways that do not challenge important foundational problems, such as accountability of industry or the empowerment of women. Surely, some groups will straddle this dichotomy, but the point remains that most poor coastal people, women in particular, are not represented in the same way USAID, Citibank, AES Corp., Cargill, or Charoen Pokphand Group are represented and served by globalization. This connectivity is important, particularly in these areas where social institutions and the State are repressive. For example, the group Down to Earth helps the Southeast Asia region by publicizing issues that groups within the region may not be able to given the degree of repression; the fact that the organization is not local gives it more degrees of freedom to work against repressive foreign state regimes.

Also, issues are combining in complex fashions. Single-issue activism is not a trait seen in many of the regional and globalizing NGOs—they are combining issues like class and gender oppression and connecting them to green politics, as is seen in the Caribbean Natural Resource Institute or the World Rainforest Movement. The NGOs are working on hunger and unequal exchange of goods at the same time they are pressing for tangible ecological preservation. These connections made by civil society might be constructed with a metaphor of a living World Ocean where social organizing is one part.

This is different from, for example, a Marxist and Hegelian orientation of resistance. Within Marx's and Hegel's concept of historical materialism, a force rises up. As it does this, the original force generates its own contradiction—its antithesis—and the two perpetually battle until the "end of history." However, in the Caribbean, for example, it is possible to argue that these NGOs were more like a self-generated, or autopoietic, response to the distress found in the region. By this I mean that they were not organized to be counterposed to the globalization and expansion of capital per se, but rather seem to be working to find a pragmatic answer that provides economic well-being that does not wreck coastal systems and inhabitants. It is even plausible to theorize that these groups emerge and continue to generate information and organization as a matter of some kind of opaque version of evolution in response to global ecological crises.

In fact, one benefit of complex systems theory is that it highlights linkage between matter, energy, and information such as human actors and conditions in ways that are compatible with and metaphorical to living systems (Capra 2002). CST can make a connection to the way these social pressures are occurring and their ecological responses because complex systems theory has a developed sense of living systems. Capra reminds us that we have a hard time defining what "life" is, even if we supposedly know it when we see it. However, as scientists and philosophers come to a better understanding of what it is, life looks like a complex system. Life is both open and closed because living systems need to exchange energy but not bleed to death (in more ways than one). Life is autopoietic and responds to disturbances and adapts to these changes in unpredictable ways. The ocean's socioecological connections found in this book are much like this, and from this characteristic I see an affirmation that human society is really part of the larger ecological matrix of life, not separate or essentially different from the rest of life on Earth.

Economic complexity is also relevant. One of the primary findings of the book is that economic globalization is connected to the regional and global ecological changes. However, from the work of Martin-Ramos (2003) and Rihani (2002), we know that economic systems are also open (to energy and change), and complex. Martin-Ramos warns that we cannot use econometric time series studies to really understand ecological economic changes. Even though this book does not use time series, the regional cases do imply a linear logic of understanding economic pressures and ecological changes. One of the links found in the cases is that as globalization increases, ecological deterioration increases—this is a linear cause-and-effect relationship, even if it is not tested in statistical terms here.

Martin-Ramos warns against such oversimplifications and suggests that it is better to measure changes after the fact, using policy as a "steering mechanism" which implies minor changes based on "cybernetic" information, or information that recalculates based on changing information. This resembles the plea for experimental human ecological policy iterated by Lee (1993) for exactly the same reasons. From this, I conclude that economy, scale, and throughput all have some expected impacts, but that there are other factors at play that are not included here but which have some influence on global sustainability; further, we can have a general understanding that economic globalization is not creating better ecosystems or improving sustainability for the World Ocean, but that its long-term effects and hidden impacts are yet unknown. It may be that these pressures will tip the world system into a transformation of economies to ones that are more sustainable.

Some problems persist, though. For example, the impacts of globalization are fast, particularly in geological time. From a macro perspective, globalization pressures have changed fisheries, coastal zones, and the thermohaline current qualities in unprecedented ways in the last thirty to forty years. Some financial interests are as short as five to ten years, as in the case of the shrimp ponds. In contrast, the fisheries may be operating on 100- to 200-year cycles (Milich 1999), making it impossible to understand the severity and long-term impact of anthropogenic change in these areas. In the case of the climate-related systems, this time frame of change is even more pronounced since the larger climate system has been relatively stable for thousands of years, and humans seem to have changed it in a matter of generations.

Likewise, development strategies look at the decade timescale, yet they are impacting coral reefs which are on cycles of thousands, or even tens of thousands, of years based on the changes found in the plants and animals in coral reefs, and their structural-skeletal changes over time. In other words, we cannot know for sure or predict the exact changes, but we can understand some of the scope and direction of these changes based on economic pressures.

Economic pressures are reducing the number of species in the ocean through fish depletion (such as through the loss of trophic complexity) and in coral (with most coral in danger over time, and much already lost to far simpler cover) in a comparatively minuscule amount of time. This indicates a very large-scale change in the system, and one that is heading more toward total equilibrium (death) than toward dynamic regeneration, and it provides a dim view of ocean sustainability. This does not mean the ocean system will end in this death; it just means that we are informing this large system and changing it more toward simplicity.

This direction also is grave for the coastal peoples of the world who are losing sustainable livelihoods, for example, in the loss of small-scale subsistence fishing and the loss of subsistence access to and use of mangroves. These sustainable livelihoods are being replaced by a globalizing economic system that is focused more on exports and growth in revenue than it is in feeding population, reducing poverty, and stabilizing ecosystems. Again, we cannot know what result this will have "in the end" because there is no necessary "end" to this process, and people will self-organize in ways we do not now understand in response to changes in the information and conditions. Nevertheless, the pressures themselves are resulting in larger income inequality, fewer cultures, less governance of corporations, and at least similar coercive suppression of civil groups.

In sum, the meaning of the global changes described in this book offer two basic interpretations of the same expectation. This expectation is catastrophic change, or a radical rearrangement of the current ecological, social, and economic orders. This undoubtedly means vast human suffering and ecological decline at the time of this change. The second interpretation from CST is that complex adaptive systems are, by their nature, creative, and new patterns of life will self-organize out of the ashes of the past system, not as a reversal to a previous pattern.

Hermeneutics

Hermeneutics is the process of understanding a "thing in itself" through critical reflection, as noted in chapter 2. I acknowledge my context and tell you, the reader, my place from which I read by telling you what I see as the "things in themselves": fisheries, coral reefs, ocean temperature and sea level changes, poverty, economic production, social institutions, and civil society within regional ocean basins, for all of which I have a personal and deep concern.

My internal benchmark for critique is the notion of sustainability marked out by Borgese (1998) through ecological health, social nonviolence, egalitarian governance, and material equity and subsistence. I believe this is a solid conceptualization of sustainability, which should be viewed as more of a process than a destination. These issues provide a depth of meaning to the words *Globalization and the World Ocean*, which I elaborate on below. First I describe how the findings of this book can be read into a theory of the global through hermeneutics, and then I describe this reading, in light of the Borgese Test for sustainability, by a traditional consideration of text and language. I then turn to a more controversial sense of hermeneutics by applying the notion of text to nature itself.

Hermeneutics encourages us to view our reading with a sense of holism. The holistic picture is made up of several messages. The sending and receiving of messages, the writing and dissemination of text, and the discussion of these texts are ways in which hermeneutics can build a theory of the global. Reasonable criteria for understanding globalization through hermeneutics would be the pervasiveness, conviction, and adoption of similar texts around the world. One way to look into such discourse in relation to ocean politics is to look to the international regional governmental regimes' texts, such as their websites and treaties, and determine if this discourse appears genuine, and to what extent it is adopted. As noted above, my internal benchmark for critique is "sustainability"; I now look to each of the regional governmental organizations to read their narratives on this area.

One thing I gather from the international regional institutions is a similar pattern of text (in treaties) and other language-based messages, such as in their websites. I find that the extent to which the discussion of sustainability, genuine or not (or in Habermasian terms, communicative or strategic), is converse to the degree of crisis in the region. In other words, the ecological crisis has been slower to set into the South Pacific, but that region's preparations and discussions of preparations for such crises are forefront concerns of the South Pacific Community Secretariat and SPREP.

146   CHAPTER 7

The protection of ecology, economy, and culture from the perspective of the South Pacific Community Secretariat and SPREP is a primary goal according to their texts, and is pervasive (a regular and consistent occurrence) in their continuing meetings, documents, and programs. CARICOM is in the middle. It has only recently organized around sustainability, and its discussions about sustainability are more shallow since they are more about the preservation of ecology and economy, but do not include culture to the same degree as in the South Pacific. They are conscious of the looming ecological changes and, of course, of the current ones, like those in Haiti. Thus, they continue to develop their communication about sustainability on these two primary levels, and the extent of their ecological crisis is not as serious as in Southeast Asia, nor as promising as in the South Pacific. Finally, I read in ASEAN that the region is concerned about sustainability only on a tertiary level. Within this level, ASEAN seems to give most of its meaning of ocean sustainability to continued fish catch and the growth of export. Thus, the development of sustainability as a notion is most shallow and least discussed in this region that has the most dreadful of sustainability troubles.

We can read more from these regional international organizations. All of these IGOs reproduce varying degrees of liberal narratives to the same extent these groups have experienced globalization. This is not an accident since a neoliberal political economy seems to go hand in hand with economic globalization; this is an observation that confirms my initial discussions of ideology in globalization, found in the introduction to the book.

IGOs in all three regions seem to be sending a consistent set of signals in reference to neoliberalism found in the actual policies at the regional and country level, though the commitment to these programs seems to vary. Some of these signals include privatization, public funding of large private-sphere enterprise related to infrastructure, laser-beam focus on monoculture exports or other "competitive advantages" such as tourism, the reduction and elimination of certain (not all) tariffs, and "austerity" programs which aim to decrease the role of the state in social, labor, and environmental programs. There are secondary narratives that discuss protecting or establishing individual freedoms such as association and speech. These narratives allow for the organization of NGOs in a more robust manner than in overtly repressive regimes; however, it is clear that these are not primary concerns. For example, the list of "Washington Consensus" demands from Western lenders for aid and loans is entirely focused on the liberalization of trade and reductions in government spending, but is not at all concerned about the liberalization of civic life, with the exception of the insistenceon private property (see, for example, John Williamson [1993], who coined the term). This is a narrative that is found to varying degrees, but is actually found throughout the three regions without exception, and I gather that the coherence of this message is consistent and infiltrating most corners of the world, and is in every sense a global discourse.

The texts of these neoliberal messages are important for natural resources because through their discourses they interpret how the policies define ocean systems. From the neoliberal focus on exports for revenue, the ocean and its biotic communities exist more and more in instrumental terms at the same time they are in increasing decline. Fish and reefs and the water column are functions within a short-term human end. These texts, such as the ADB Policy on Fisheries (1997), are now beginning to connect the natural resource systems to other conditions of human security. This is an important improvement, but ADB's policies of funding and appreciation do not raise the importance of biotic life in the ocean beyond a means to an end, nor do they change the focus of bank loans aimed at simple economic growth. This kind of meaning for ocean systems, and for nature in general, is not sustainable (Ridgeway 1996) because it encourages exploitation based on economic efficiency, not sufficiency (Princen 2003).

This is a finding from hermeneutics that is worth reiterating: efficiency alone is not sustainable. Efficient use of resources is necessary, but not sufficient to bring scarce resources into the future because it lacks an ethic for structure and system, and is based mostly on Western values of individualism—a credo that has largely delivered us to the unsustainable point we now find ourselves (Oelschlaeger 2001).

Neoliberal economism, in any of its stages, sees nature only as a resource base, not as something that is life, contains life, or reflects our own life. It cannot recognize the obligation and committed relationship to nature that is required for a healthy and sustainable world. One reason for this may be liberalism's demand for a neutral state that grants protections only to individual humans, not to ecosystems or organisms within ecosystems. If protections through policy are granted, they are subject to revision and removal when the political tide turns and there is no out-front commitment to sustainability as such, as there is for the utility of individual choices and consumption. Planning for the future with sustainability as a prima facie goal is not permitted by the standing rules of liberalism (Ridgeway 1996). Thus, we can see that one insight of hermeneutics for the World Ocean is that liberalism and its transformations, like neoliberalism, will consistently end up with very wide gaps in relation to the Borgese Test because the meaning of sustainability in this ideology is itself thin and therefore insufficient.

Worse, neoliberalism's view that nature is a commodity that *needs* to be exploited or is otherwise wasted in light of human needs because the ecological community does not factor into its immediate close concerns is patently unsustainable and undesirable from the system view. In this way, liberalism is anthropocentric, separating humans from nature, and this too is simply not a sustainable basis for action since it permits privilege and abuse to both human and natural communities (Oelschlaeger 2001; Hawkins 2002). In conclusion, the same text is being written in all of these regions with different degrees of insistence and local modification, and it appears to be emerging from the centers of power globally through North America and Europe, and from centers of power regionally through the United States, Australia, and Japan.

I now turn to a more provocative proposition: human communities are not the only ones writing down messages for us to understand. In order to approach this admittedly controversial subject, I turn to the ecosemiotics and a special issue of *Sign Systems Studies* which was dedicated to the area of study.

According to Winfried Nöth (2001), ecosemiotics tries to understand how "the relationship between humans and animals and their *umwelt* [environment] . . . is mediated by signs" (71). Communication is the sign process between a sender and a receiver, but does not need to take place only between humans. He argues that communication occurs between "all other organisms in the whole biosphere" (72). Nöth notes that historically our idea of communication was centered on human language, so that nature was capable only of being a referent or the content substance of this language. This would explain one reason why our notions of hermeneutics typically do not go beyond human language messages and text.

I propose here that there are other kinds of messages to be interpreted—those from ecology and the organisms of Nöth. Further, these signs are not necessarily separate from us, the human receiver. This is in keeping with the notion of synechism, or the continuity between mind and matter, and is contrary to Cartesian dualism, which separates mind and matter by placing them in opposition to each other where the mind understands a distinct matter. Importantly, creating an opposition/dichotomy between mind and matter allows for powerful and practical decisions that allow natural-resource exploitation in unsustainable ways through the well-known and oft-discussed separation between culture (mind) and nature (matter) with an implied hierarchy favoring the former (Nöth 2001).

Synechism permits a unique kind of epistemological position. Instead of objective observation's informing the subject about the facts, or the converse—that there are no discernible objects or subjects as postulated by postmodernism—synechism allows for the building and interpreting of signs and messages from real existing nodes in the world. Therefore, it is possible to build a genuine understanding of the world, even if this understanding is not unilateral. In fact, it is because nature and culture/humanity are not excluded from one another or dialectically opposed that a more whole understanding of nature is attainable.

Because we are not conceptually alienated from nature, we have a connection that informs our dynamic understandings of the world, and knowledge building becomes possible. At the same time, this kind of interpretation requires our own admission that any kind of Absolute Truth is unobtainable and that our informed connection occurs within a subjective context. In short, nature is not alienated from mind and culture, and mind and culture is not alienated from nature. Imposing such alienation may, in fact, be the very source of our environmental problems (Oelschlaeger 2001). The meaning of this position is that there is purpose in some environmental changes, and that our own mind understands its world partly from evolving from within nature itself (Nöth 2001). We can build knowledge about the World Ocean because we are a part of it.

From this point of view, the evolution and communication between mind and community and nature can be read as a set of negotiations. Alf Hornborg (2001) believes that human-environmental relations can be seen as a set of signs which are passed through the continuity of environment and humanity. He notes that in Amazonia, early signs between the nonhuman environment and human culture were made through oral and sensory messages that were transmitted directly between human-forest systems without being mitigated by other mediums.

However, over time, economic sign systems replaced this direct set of negotiations, distancing the human communities from the messages of the Amazon and decontextualizing humans from the knowledge systems that were embodied in the more direct relations. When we expand Hornborg's thesis to this project, we can view the deterioration of the World Ocean biological and human systems as a problem of communication between the two, further indicting the neoliberal instrumental reason that compounds and reproduces this distance and delegitimizes direct communication with the full range of ocean human ecology. "Listening" to the ocean is nonsensical and ridiculous from instrumental and economistic positions because the ocean and its parts are objects, not life communicating or worthy of nonexploitive relationships. This may be why some within the neoliberal persuasion continue to see the world as "not without problems, but on almost all accounts, things are going better and they are likely to continue to do so into the future" (Lomborg 2001, xxiii). In other words, the communicative signs regarding nature are distorted, decontextualized, and in some way fetishized (falsely accounted) through the economic sign systems that are most important to this set of anthropocentric narratives. If our narratives are based solely on human wants and needs, we can see how the end of direct communication with nature also ends serious negotiations and imposes unilateral human demands regardless of larger ecosystem needs. Hermeneutically, our crises in the World

Ocean can be viewed as the end of reciprocity and communicative action between human and natural agents.

The situation is not yet as Borgese had hoped—where the ocean would bring disparate international communities together for peaceful purposes based on the metaphor and messages from the ocean itself. If anything, the ocean continues mostly to bring business together between tourism, fisheries, and container ships through the current global focus on liberalizing trade and decreasing governance of this trade as well as social, labor, and environmental problems. However, there are some discourses that indicate hope. For example, the Tamil Nadu Gram Swaraj Movement (*swaraj* is radically decentralized political economy), which is located in India's coastal communities and opposes the farming of shrimp, has made important advances. This movement is directly informed by Gandhi's political economic visions of nonviolence and is a sustainable-development movement organized in opposition to the globalization of centralized economic exports and the structural adjustments imposed by the World Bank and IMF, which have directly hurt the poor, women, and other marginalized groups (Rigby 1997). This movement has spread some throughout Southeast Asia, is connected to the global Third World Network, and has coordinated with the United States–based Mangrove Action Project (Rigby 1997). Nonetheless, these advances of grassroots local efforts making the global political scene are exceptions, and many of the regimes continue to struggle with authoritarianism.

Critical Theory

Critical theory can also provide us with a theory of the global and a sense of ocean sustainability, which I now discuss. A critical theory of the global relates to a political-economic network. From the Frankfurt School of Critical Theory, we understand that there is a concerted effort to centralize control over the masses of people as well as nature, organized through the capitalist system and the semi-autonomous state which they deride. The capitalist system reproduces itself across geographic areas in order to expand production and consumption through a world capitalist system; this process concentrates power with economic elites that hold most of the capital. Thus, we see from trading flows whether or not there is a connected global political economy moving surplus economic and ecological value from the periphery (poor countries) to the affluent commercial hubs or centers.

Some of this is reflected by this study. There is a global flow of fisheries, coral reef commodities, and forcing that is class related. Affluent centers of power determine the flow of fishery goods and coral reef commodities like aquarium fish, and climate change forcing is from emissions almost entirely from wealthy countries, which have the longest historical contributions to carbon emissions and have built hydrocarbon-based economies. The effects of climate change seem to be most intense for small island states which are not part of this forcing in relative terms and which are normally poor.

In each region there is a very clear hub defined by the volume of FDI, just as critical theory would suspect. In the Caribbean it is the United States; in Southeast Asia it is China (neither this nor most other studies defines Japan as belonging to the region); and in the South Pacific it is Australia. The rate of FDI change in hubs appears relative to the level of commitment to sustainability in the region, except with climate change that is skewed by the United States' 30 percent of world carbon emissions and appears to correlate with absolute FDI in the hub. Comparing hubs, we see that China experienced the highest rate of change of FDI, at roughly three times the rate change in FDI found in the United States; this is followed by Australia, and the rate of change in each hub correlates with the rate of change found in the region as a whole. This probably means that the rapid FDI change in the hubs is partially responsible for guiding regional economic flows and therefore partially determining resource impacts and exploitation in addition to the way governance responds to these problems. If this is true, then both the Marxist and liberal expectation that political economy forms the "base" for the rest of sociopolitical and ecological relations has a great deal of credibility when it comes to the World Ocean.

From critical theory, our central concerns are political economy and knowledge control. Critical theory is fundamentally skeptical of unifying and centralizing since such centralization robs power from the underprivileged. One concern in particular is the distancing or "distanciation" of ecologies mentioned in the introduction. Therefore, when we look at the picture of sustainability in the World Ocean, it is important to ask who pays the ecological costs of economic expansion, and if knowledge is controlled through a similarly hierarchical system.

Fisheries follow this model in our study, but only in a qualified way. Most of the fish exported (at least 70 percent) go to affluent nations, and many of these exports come from developing nations, who have received loans from centers of financial power controlled by these same countries, like the World Bank. Also, the global flow of fish threatens the livelihoods of small-scale fishers. These are people who are often the poorest of the world's poor, may make up as much as 90 percent of the world's fishers, and catch as much as 45 percent of the world's fish, mostly for direct human consumption (McGoodwin 1990). They are increasingly in jeopardy from industrial fleets which send most of their fish to affluent consumers. This means more food insecurity for these particular groups

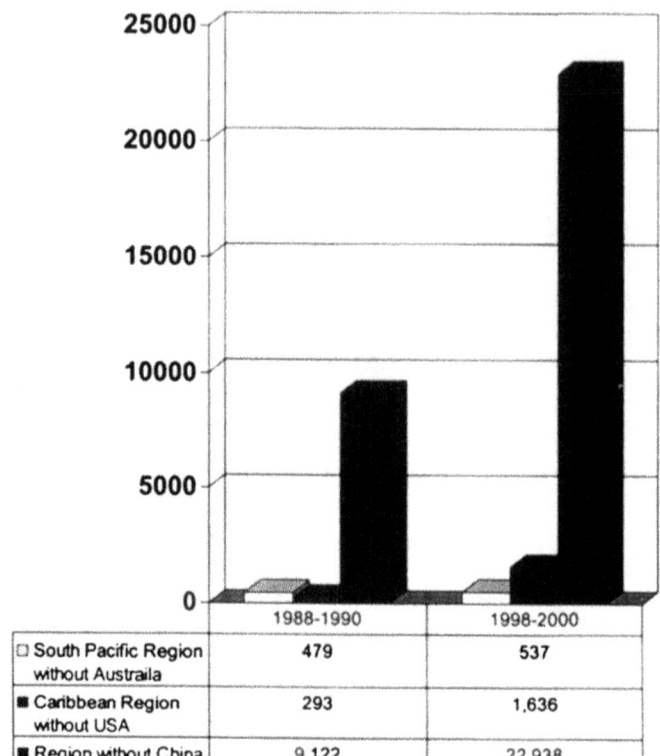

Figure 7.1. Average Regional Foreign Direct Investment (FDI)
Source: Adapted from World Resource Institute 2003a.

who in one region have generations of tenure that creates deep wells of ecological knowledge, and who typically look at marine ecology holistically instead of as a set of single species (McGoodwin 1990).

Global capital in fisheries is severely threatening traditional fishers and their more holistic knowledge of Southeast Asia. However, the Caribbean is a contrary case that offers a caveat to overgeneralizing about small-scale fishers. Here it is not centralized knowledge systems and centers of power depleting the snapper and grouper populations; rather, it is the local small-scale fishers along with the global tourist industry. In Jamaica, this may be indirectly related to the worsened poverty affiliated with SAP loans. However, global TNC fleets are not inundating the waters of the Caribbean basin as they are in Southeast Asia or in the South Pacific. The global centers of capital are more interested in the large catches of

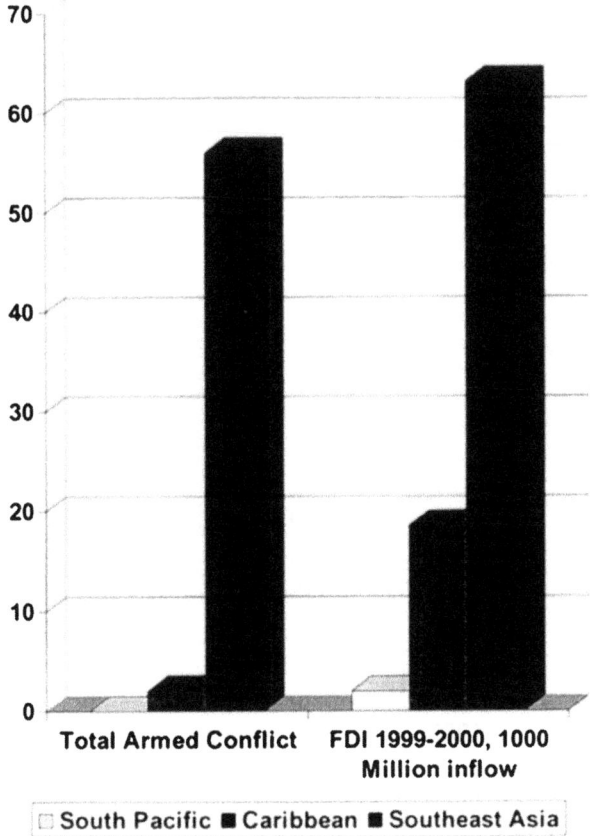

Figure 7.2. Armed Conflict and Foreign Direct Investment (FDI) between Regions
Source: World Resource Institute 2003a; Eriksson, Wallensteen, and Sollenberg 2003.

tuna than in the small catches of grouper; therefore, only some of the World Ocean fisheries are globalized. Lower nutrient levels in the region may be a limiting factor that has determined the locus of world fishing activity.

From a critical theory perspective, globalizing economic forces are expected to widen the gap between poor and affluent countries (Kim and Shin 2002). Since I include international economic equality as a part of sustainability, globalization that widens the gap between these countries, creates, holds steady, or improves only in a token fashion national poverty or bloc poverty (poverty held between regional peers) is not sustainable. Indeed, the ratio of the share of wealth

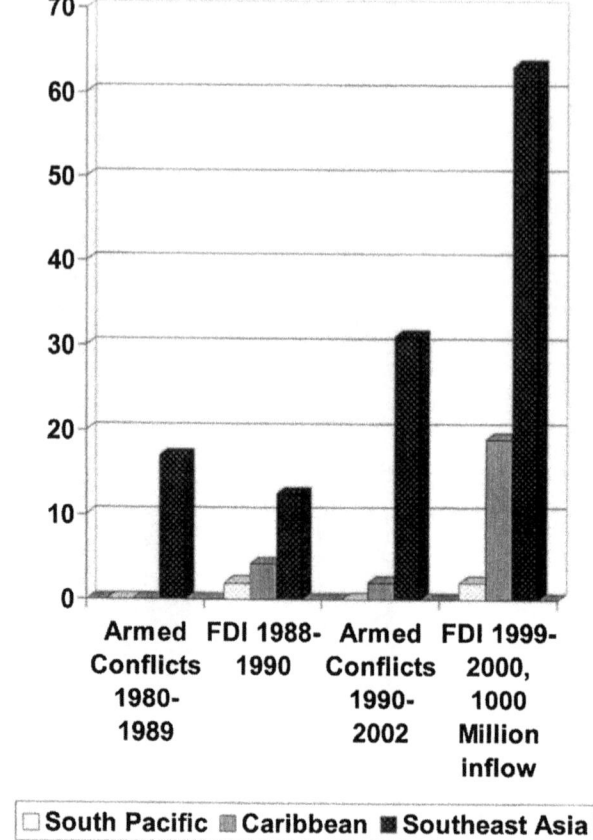

Figure 7.3. Armed Conflicts and Foreign Direct Investment (FDI) across Regions, 1980–2002
Source: Adapted from World Resource Institute 2003a; Eriksson, Wallensteen, and Sollenberg 2003.

held by the top and bottom wealthiest people in the world has doubled from 30:1 to 61:1 (Castells 2001). Contrary to the claims of neoliberals, poverty has widened (Hunter Wade 2004).

Knowledge production is also important. Knowledge provides a basis for controlling a resource. For example, under the Law of the Sea, the maximum sustained yield for a fishery in an exclusive economic zone is supposed to be scientifically determined by the respective country in order to determine the extent of total allowable catch by that country and by foreign fleets. Thus, fishery population science will be a large determinant power in each country that has rich fisheries. What kind of science is acknowledged and respected? Within international organizations, Western science dominates, and is used instrumentally to determine export levels, mostly to the United States, the EU, and Japan.

However, critical theory does not seem to help us so much in this case, because the nature of the knowledge production does not boil down into a simple stew of cause and effect. In the South Pacific, even though the fishing fleets are foreign, the knowledge production has global and regional origins (it is mixed) in all regions. Some of the leading fishery work in the world is being done within the South Pacific region, for example, by Patrick Lehody; and in Southeast Asia, by Daniel Pauly (a professor at a Canadian university but who has played a key role in the International Center for Living Aquatic Resources Management, ICLARM, and is now at the WorldFish Center in Malaysia). Further, it seems that this important knowledge is not generated solely for the interest of Western countries, industrial affluent countries, or even centers of capital. This work is more meaningful than that, and there are important ramifications of this work which, as I noted in the above section, will have some kind of impact on improved sustainability.

Further, not all of the knowledge being produced is by the formal rigging of IGOs or outside funding. Outside of international formal institutions, informal science and traditional knowledge have an important role to play, even if this influence has been marginalized from the first waves of global imperialism.

All of the regions have pockets of comanagement experiments and efforts in play, and these efforts all show promise of being more democratic, more economically equitable, and more ecologically sustainable. Each region has an indigenous history that can provide information and a legacy that may become more and more emphasized, and I believe that these histories should be emphasized even more than they currently are since longtime memory of specific human ecology is of profound importance. However, we should not be surprised that these small projects are not the majority in any of the regions, and critical theory does grant us the vision to see that along with globalization of commerce came a spread of Western science. Indigenous knowledge systems are alive, but they are by no means the mainstream anywhere in the regional institutions. From a critical-theory point of view, this means we should expect that instrumental reason will become even more entrenched and nature will more and more be treated like an object of trade than something that is alive and with its own agency unless the modern institutions can adopt noninstrumental means of legitimating knowledge.

Conclusion

All of these theories provide the possibility for developing an explicit theory of the global, and an intricate notion of sustainability. In these concluding remarks,

I draw these three theoretical outlooks together in order to see what kind of consilience there is in the findings of this book in relation to ocean trends.

This will provide an additional layer of global theory—what understandings do all of these theories actually share? After concluding comments about a composite global theory, I comment on composite conclusions about the sustainability of the World Ocean regarding the areas of concern described in the regional chapters including the history of globalization and a note about the role of the United States, fisheries, biological diversity (coral reef and climate change issues included), and issues of governance.

Observations shared between each outlook about what is "global" include ideas of connections and relationships. The extension of relationships between systems—either ecological, communicative, political, economic, or those of mind and science—are the keys to knowing if something is global or globalizing. Relationships that are growing across geographic regions are globalizing in extensity, while relationships that are growing in depth, frequency, and commitment across geographic regions are growing in intensity. Relations that span the entire universe of possibilities are at the maximal scale of globalization. All of the above theories are concerned with these qualities.

Looking to the history of globalization, we observe that each region experienced the first wave of globalization through the extension of imperial force from the West made possible by the World Ocean-as-highway. Most of the globalization in trade, knowledge building, and NGO development has at least some connections to the West. Clearly, this history of globalization has not stolen anyone's complete reservoir of agency, but just as clearly, this history of globalization has created a vast network and structure that informs how the current relationships of commerce and science are accomplished, and to a lesser extent how NGOs develop. One lesson from this history is that commerce and material power are opportunistic. The current wave of globalization appears to be no different in this sense when seen in the United States, which is a leading agent in this current wave and is using its material power in ways that have a mostly deleterious impact on sustainability. This situation deserves special attention.

Regarding the major ecological changes around the world, the United States is a leading force in the cruise ship industry, supplying a bulk of the passengers and a majority of the consumers for the drug trade that plagues the Caribbean. The United States is also the world's leading emitter of carbon, the most important human-related climate change driver. This is literally killing coral reefs in large percentages, and is an act which has direct impacts on fish nurseries and coastal storm and sea level rise protections for people who typically do not contribute to global carbon emissions in commensurable ways. The fact that the

United States is also the primary aquarium-fish consumer is also a major factor in the loss of coral reefs. Further, the United States is one of the top two consumers of shrimp. The other is Japan, another important global center, and this is a factor in unsustainable coastal development, social policy, and commons management. All of these factors also have local counterpart agents, but there is no doubt that the United States is among the more powerful agents in ocean matters—it is, for example, the only globally forward deployed navy (Jacques and Smith 2003)—and it is mostly driving ecological conditions toward undesirable and potentially irreversible changes.

Through complex systems theory, the United States can be seen as an attractor of information and structure through its own matrix of commerce and material power, which then is significant in creating the system itself. In order to change the global capitalist system, the relationships with the United States and the rest of the world will have to change. CST demands that ultimately, if the United States is in part creating a stable system, the system will become more complex because more and more nodes will be allowed to gain a foothold. Globalization of commerce is increasing the diversity of members in the global market, though there is a measurable concentration of firms at the top, even if the quality of these relations is suspect.

However, when we connect the commercial system to the ecological one, a more complex biological world is not apparent. Therefore, if we define the Earth system as being the sum of the commercial/economic, social, and all ecological systems together, the loss of diversity in the latter two indicates an unstable larger system. In other words, the United States may be stabilizing the global economic subsystem, but this effort is undermining other parts of the Earth that will ultimately disrupt these very efforts.

This would also mean that changes are likely on the way for the role of the United States in the larger sociopolitical world. The United States cannot be expected to continue to maintain its position of relative hegemony if this very unipolar position destabilizes vast social and ecological patterns around it; this structure has already begun to unravel in Southeast Asia (Beeson 2004). From a hermeneutic perspective, the role of the United States and its hegemonic power is one that interferes with the messages from other agents and the ecological world. The United States itself has the power to consume other countries' and regions' resources while distancing itself from local consequences. Also, given its use of instrumental reason and ethics in relation to nature, the United States has undoubtedly created numerous intermediary relations with nature so that the direct signs from nature, and its limits, are hidden. I assume that this kind of communication block is another way that limits the viability of U.S. hegemony and its future security. Similarly, from a critical theory perspective, such as according to Wallerstein's (1989) "world system theory," where the hegemonic powers order a coherent and single capitalist system, this power historically operates in phases where hegemons overextend themselves so much that they devour their own power base and create their own disintegration, opening up the way for a new hegemon. Indeed, as much as this perspective is informed by the concept of historical material dialectic, the creation of a hegemonic order creates and embeds its own antithesis, and the role of civil society and other nations and forces will be to undermine this material power in the world over time through counterhegemonic resistance. Thus, through all of these theories, it is possible to see that singular agents of unsustainable systems create their own means of insecurity in the same way that they create insecurity for others.

Pragmatic ramifications of this loss are the changes that are occurring in fisheries, and therefore in food security for the world. Overfishing has been shown to affect fisheries in nonlinear ways, indicating that the lessons from complex systems theory may be important. For example, if the Atlantic cod is any indication, fisheries can sustain themselves in the face of mounting pressures until they approach some "cliff" of permanent decline and perhaps decimation. Given that about three-quarters of the world's fisheries are facing such pressures, we should view this potential with the utmost gravity. The language of the ocean continues to tell us through fairly clear signs that this limit is real—fish are becoming harder and harder to catch, and the kinds of fish caught are increasingly found lower on the food chain. The world's poor, even when their commons are not being enclosed for private interests, are going to feel the first human burden because they depend on this fish for more basic survival than affluent consumers who have other choices. That fish is simply becoming more expensive and harder to attain is one example of how our depletion of fisheries will further threaten the security (overall well-being) of the most vulnerable people.

Ecological and social diversity is becoming simplified at the same time, and should not be seen as accidental, but rather as a function of structural pressures creating global patterns, demonstrated by loss of higher trophic levels of fish, the loss of coral reef around the world through climate changes and unsustainable coastal development, and the loss of mangroves and coastal forest and grasslands in addition to the losses of indigenous cultures, languages, and lifestyles that have persisted for eons (which in itself says something about their sustainability). Complex systems theory sees this as unsustainable in relation to the future options systems can take; hermeneutics sees this as unsustainable because it is a sign of a sincere loss of meaning in the world; and critical theory sees this loss of social and ecological diversity as an unsustainable concentration of power that enables abuse and exploitation of nature and nonhuman nature. In all cases, humanity is diminished by such losses because we are a part of these threatened spheres of the World Ocean. Between the three perspectives, then, ocean communities are reducing their options for future pathways, losing depth and meaning in addition to the relative power to resist such trends.

Gender inequality pervades each of the regions with only a little variation, apparently found mostly on the local level. Women are disempowered in each of the regions, and this has important implications for sustainability according to each of the three theories. In CST, the suppression of nodes in the system will again have a negative effect on available options in future systems. In hermeneutics, a reading of the whole social system sees that welfare is not improving, and key conditions indicate that over 50 percent of the world's population experience a disproportionate share of violence, poverty, and ecological problems in their labor in the household and in the workforce. Sustainability is implausible for only one gender, and these conditions indicate that the meaning of sustainability very often overlooks the condition of women in society. Even as I make this note, I admit that the conditions of women have not been the focus of this study and I can see that this area requires more research and theorizing.

From what little attention I have paid this issue, it is clear that information and knowledge are organized without a gender component, leaving the lives of women unaddressed and mostly silent, a state that is a prerequisite for the institutionalization of social hierarchies (Enloe 1990).

So long as current power relations and governance structures in and out of civil society remain the same and rely on the continued silent work and suppression of women, none of the improvements in sustainability will matter much, and half of the world's lifestyles will be relying on the other half's work. In the end, this is representative of the different levels of hierarchy that are experienced in civil society and in the organization of government that Gandhi and Borgese warn against. So long as society looks more like a pyramid with the apex resting on the conditions of the base, the World Ocean communities will not be sustainable.

Conclusion 8

Ever-Widening Oceanic Circles

WELL, NOT REALLY. Elisabeth Mann Borgese thought of the World Ocean as a medium to bring humanity into nonviolent, sustainable relations. When I began this project, I had hoped to find this process starting because of the move by some areas to create formal regions; while I am optimistic about the possibilities for such relations, I do not see them being developed in major. Still, there are two stories found in this book regarding widening ocean circles. The first story is that even since the colonial period, concern for problems about the ocean is not bringing and has not brought relatively modern and modern nation-states together for peaceful purposes.

With the advent of the Fishery Stock Agreement, there were signs of this possibility. This groundbreaking agreement broke with the historical and powerful demands of mare liberum and instituted the possibility of enforced international law through regional international relations based on precautionary science and a concern for sustainability of transboundary fisheries. I had hoped to find these oceanic concerns drawing the regions together as a basis for a new foundation for sustainability in our globalizing world. This is not happening. Material power in arms, commerce, commodity, export, and ideology are the basis for most of the international relations at the regional level found in this book, while sustainability in two of the three regions is of secondary concern. In the South Pacific, there is hope for Borgese's oceanic circles. Concern in the region for good governance, knowledge building, and limits on the exploitation of natural resources are signs of some promise.

Barring some reversal in politics, as long as a focus is maintained on removing governance from industry and trade worldwide (while keeping the State heavily involved in civil society), the ability for citizens in poor countries to redress problems will not be strong enough to turn the tide against unsustainable trends. Kai Lee (1993) writes about this issue even more broadly: "Surely one of the messages of the twentieth century to posterity will be that our science and technology persistently outran our ability to govern our expanding capacity to change the world and ourselves" (4).

Failure in evolving sustainable processes has had disastrous results in the past. Sing Chew (2001) has done research dating back five thousand years that suggests that ecological degradation over long periods of time produced by historic hegemonic trade expansion has produced global systemic crises and collapse which has led to "Dark Ages," or ecological crisis. Chew looks to the current world capitalist system in conjunction with increased ecological crisis and warns, "The indicators of ecological stress and distress, and the climatological changes, do suggest that we might be approaching another Dark Age, if we are not in it already" (Chew 2002, 353).

Globalizing economic intensity seems to be a driving force behind many of these key ecological changes, and I have traced some of the specific lineages of these processes, directly tying globalizing market and financial forces to these problems. In areas where the intensity market forces changed the most, fisheries and coastal ecosystems headed proportionally toward decline. Absolute capital was associated with higher carbon emissions. Changes in the World Ocean in these areas limit the regional ability to choose among several options, and key ecological systems have been potentially irreversibly changed.

Market forces are the most developed connections between industries, countries, regions, and institutions in this study. Planning and knowledge building in natural-resource protection and human subsistence are present and building strength, but are fragmented and have less power in the world today than export, lending, credit, austerity impositions as in the Washington Consensus, deregulation, industry consolidation and privatization, and profit.

World Ocean Security as Global Human Security

In this section I discuss ocean security as human security. By security, I mean holistic well-being and health and the absence of threats to that well-being. This is the second, more interesting story about Borgese's hope for oceanic circles. It is more subtle, but it is evident here—regions that are more sustainable are more peaceful, and this sustainability underlies several layers of other kinds of security. In other words, how we live with and regard the World Ocean has something to do with how we live with and regard ourselves on a global level. According to Eriksson, Wallensteen, and Sollenberg (2003, see figures 7.1 and 7.2), the most sustainable region, the South Pacific, had zero—none—armed conflicts in the last fifty years or so; the Caribbean had a few; and Southeast Asia had an overwhelming amount of conflict, almost all of which have happened since the globalization boom of the 1980s, with a majority of the conflicts (thirty) happening in the last ten years as globalization became more intense. This does not establish causation nor does it address the clear evidence that international trade (along with democratic regimes) decreases interstate war (Oneal and Russett 1997), but it does indicate that armed conflict has some relationship with sustainability that beckons us for further study. This is compatible with the "green peace" I statistically tested under narrow conditions for the year 2002 out of 142 countries. More sustainable countries were more peaceful than less sustainable countries by 99 percent (Jacques 2003).

The differences among the regions' military security really surprised me, but it reminds me of the fish off the Grand Banks. They suffered the hooks and nets of increasing pressure for some time, and then not gradually, but dramatically, fell off. I wonder if the social ability to avoid violence and maintain a stable region is like this—if a region can fight off the pressures for some time, and then, as seen in Southeast Asia, this ability is worn away. Certainly, in order to help assuage the misery of this region and protect other regions from this same fate, social and physical scientists need to come to a better understanding of the nature of our human security as it relates to ecology.

This kind of re-visioning is not new thinking. First, green political thinkers, across a variety of perspectives, share the common perspective that how humanity treats and regards nature is akin to how we treat and regard ourselves (Jacques 2003). This is borne out in this study as it relates to the ocean environment.

Second, since the 1970s, political scientists have been rethinking the concept of "security" to include environmental factors (Tuchman-Matthews 1989). Both Katrina Rogers (1997) and Jon Barnett (2001) have identified the more important strain in this new thinking as "ecological security." In one sense, "ecological" means "universal" and "comprehensive." In a more direct sense, it means that nature provides human groups with vital goods and services necessary to survive. Protecting these services in the World Ocean protects our ability to live as well as produce, consume, and subsist. Deterioration of the World Ocean does the opposite.

This way of thinking about security reflects the difference between what Elise Boulding has described as "positive" and "negative" peace (2000). From a perspective of ecological security, the findings of this book indicate that unsustainable processes and conditions are leading directly to tangible insecurities. Let's recap some of the story lines in this book, now with the understanding that sustainability relates to human security.

Ocean Sustainability, Ocean Security

Fisheries are declining, coral reefs are dying, and mangroves, sea grasses, and the coastal zone are being removed and killed. None of these ecologies in any of these regions are generally improving. The ocean will be able to recover from some of these changes. Not all of the coral mortality will be permanent, and some of the animals that inhabit these structures will come back; global warming may actually help some mangroves grow in new places; and as many as 25 percent of the world fish populations are not overfished (if FAO statistics are correct). Further, those populations that are overfished but not depleted may be able to recover their populations if pressure is taken off before this happens, as noted by complex systems theory in addition to linear fish science.

However, some of these trends are irreversible and global. The changes found in local/regional coral reefs will have global impacts in their ability to hold key areas of life together in the tropical seas, at the same time that mangrove fishery nurseries are disappearing and fishing pressures suffer from fifty years of continuous increase in capacity and effort. It is no wonder that the marine trophic levels all around the world are becoming more simple and flat, indicating a declining world gene pool. Globalization of neoliberal economic policies and the extension of big capital are part of this problem—the more intensely a region experiences globalization, the less sustainable its ocean systems are.

Industrial nations must take responsibility for a large degree of these changes. Tuna and shrimp production are organized by Northern funding and go to Northern consumers; tourists in the Caribbean are almost entirely from the North; and of course there is climate change. Industrialized countries are industrialized because they converted their economies from the solar energy base in agriculture to finite hydrocarbon stocks. They have changed and are changing the nature of the water column, as demonstrated through the dramatic absorption of carbon within the oceans, the increased sea levels, and the global temperature increase. Thus, to the extent that the United States leads global carbon emissions, it leads global climate forcing and creates insecurity for the rest of the world, in particular the small island nations and coastal areas, such as in Bangladesh. However, this should not indicate a free pass for such industrializing countries as China and India, which also clearly need to take responsibility for their emissions.

All of these factors indicate that the World Ocean is changing in unpredictable *but consistent* ways. The depletion of fish resources and the increased competition for fish may provide the grounds for conflict, but more importantly, the loss of ecological goods in services in the coastal zones and in fisheries undermines the basis for subsistence for many thousands of poor people who have used these ocean commons for millennia. Their security is most damaged. Further, among these groups, women and ethnic minorities, as well as geographically marginalized people such as the so-called boat people of Southeast Asia, may be the most insecure of all.

However, this insecurity will not be contained for very much longer, even if it focuses on and starts with the less powerful. This insecurity will climb up the social ladder. The changes to these ocean ecosystems are global and will have some kind of impact on every person around the world. Even corporations will not benefit in the long run from declining ecosystems since there will be less and less to exploit. Perhaps this is what will have to happen to realize Borgese's ocean circles.

I conclude this book with a final thought and then a rhetorical question. The modern relationship with the World Ocean is not sustainable. The ocean system is deteriorating and structural elements of the ocean are changing globally. This is not just a loss of security, but a loss of meaning and identity for humanity because we are a part of the ocean—we depend on and gain life from the ocean, and whole societies put the global ocean at the center of their life. In many ways, we erode away a base of our humanity when we choose to undermine a cosmically unique life system like the World Ocean for instrumental, short-term, and economistic purposes.

I finish this book as I look out into the Pacific from Hawaii. The vast scope and eternal rhythm and connectedness—the peace—of the ocean settles deep within my chest. I know Borgese was right—that the ocean can be a model for human organization if we listen to the ocean itself. But, we are alienated from one another and the ocean—indeed, nature—unlike more sustainable indigenous peoples, as in the precontact South Pacific. It is unlikely we could (or should) return to this exact type of social organization. So my question is this: How can a postmodern world overcome this alienation and turn the current global connections into an altogether new and sustainable vision of development? What will it mean to live on Earth and be human if we do not?

References

Acharya, Amitov. 2003. "Democratization and the Prospects for a Participatory Regionalism in Southeast Asia." *Third World Quarterly* 24:2, 375–90.
Adams, Chris. 2001. "The Asian Development Bank, Capital Flows, and the Privatization and Infrastructure Projects in the South." In *Profiting from Poverty: The ADB, Private Sector, and Development in Asia*, 11–22. Bangkok: Focus on the Global South, at www.focusweb.org (accessed June 13, 2004).
Adams, T., and J. Majkowski. 1997. "South Pacific Islands." *Review of the State of the World Fisheries*, FAO Circular No. 920 FIRM/C920 (En), at www.fao.org/docrep/003/w4248e/w4248e31.htm#AREAX7 (accessed January 28, 2004).
Ahmad, Nilufar. 1995. "Battling the World Bank." *Multinational Monitor*, at www.multinationalmonitor.org/hyper/issues/1992/10/mm1092.html (accessed June 19, 2004).
Ali Khan, Tariq Masood, et al. 2002. "Sea Level Variations and Geomorphological Changes in the Coastal Belt of Pakistan." *Marine Geodesy* 25:^{1}/$_2$, 159–74.
Alonso, Irma. 2002. "Social Conditions in the Caribbean." In *Caribbean Economies in the Twenty-first Century*, ed. I. Alonso. Gainesville: University Press of Florida.
Arctic Climate Impact Assessment (ACIA). 2004. *Impacts of a Warming Arctic: Arctic Climate Impact Assessment*. Cambridge: Cambridge University Press.
Arias-González, Ernesto, et al. 2004. "Trophic Models for Investigation of Fishing Effect on Coral Reef Ecosystems." *Ecological Modeling* 172:2–4, 197–212.
Arnold, Ron, and Alan Gottlieb. 1994. *Trashing the Economy: How Runaway Environmentalism is Wrecking America*. Bellevue, WA: Merrill Press.
Ascher, William, and Natalia Mirovitskaya, eds. 2000. *The Caspian Sea: A Quest for Environmental Security*. Boston: Kluwer Academic Press.
Asian Development Bank. 1995. News Release, at www.adb.org/Documents/News/1995/nr1995129.asp (accessed June 19, 2004).
———. 1997. "The Bank's Policy on Fisheries," at www.adb.org/Documents/Policies/Fisheries/default.asp (accessed July 25, 2004).
———. 2002a. "United States and Shareholding and Voting Power," at www.adb.org/NARO/usa.pdf (accessed June 18, 2004).
———. 2002b. "Project Completion Report on the Fisheries Development Project in the Federated States of Micronesia." PCR: FSM 24267, at www.adb.org/Documents/PCRs/FSM/pcr_fsm_24267.pdf (accessed June 19, 2004).

———. 2004. "Coral Reef Rehabilitation and Management" at www.adb.org/Documents/Profiles/LOAN/29313013.ASP (accessed June 19, 2004).

ASEAN Regional Centre for Biodiversity Conservation (ARCBC). 2002. "Marine Protected Areas in Southeast Asia," at arcbc.org/arcbcweb/publications/mpa.htm (accessed November 16, 2004).

Bailey, Conner, Dean Cycon, and Michael Morris. 1986. "Fisheries Development in the Third World: The Role of International Agencies." *World Development* 14:10–11, 1269–75.

Banks, Helene, and Nathaniel Bindoff. 2003. "Comparison of Observed Temperature and Salinity Changes in the Indo-Pacific with Results from the Coupled Climate Model HadCM3: Processes and Mechanisms." *Journal of Climate* 16, 156–66.

Barker, David Read. 2002. "Biodiversity in the Wider Caribbean Region." *Review of European Community and International Environmental Law* 11:1, 74–83.

Barnett, Jon. 2001. *The Meaning of Environmental Security: Ecological Politics and Policy in a New Security Era.* New York: Zed Books.

Bateson, Gregory. 1979. *Mind and Nature: A Necessary Unity.* New York: Dutton.

Beeson, Mark. 2003. "ASEAN Plus Three and the Rise of Reactionary Regionalism." *Contemporary Southeast Asia* 25:2, 251–69.

———. 2004. "U.S. Hegemony and Southeast Asia: The Impact of, and the Limits to, U.S. Power and Influence." *Critical Asian Studies* 36:3, 445–62.

Benchley, Peter. 2002. "Cuba Reefs a Last Caribbean Refuge." *National Geographic* 201:2. Also available online via Academic Search Premier (accessed May 17, 2004).

Bende-Nabende, Anthony. 2002. *Globalization, FDI, Regional Integration and Sustainable Development: Theory, Evidence, and Policy.* Burlington, VT: Ashgate Publishers.

Bengwayan, Michael. 2002. "Philippines Fast Becoming the Land of the Dodo." *Asian Observer*, August 22, at www.asiaobserver.com/Philippines/Philippines-story6.htm (accessed June 19, 2004).

Benn, Dennis. 1987. *The Growth and Development of Political Ideas in the Caribbean.* Kingston, Jamaica: Institute of Social and Economic Research.

Bestor, Theodore. 2000. "How Sushi went Global." *Foreign Policy* 121: 54–63.

Bigg, Grant. 2003. *The Oceans and Climate.* 2nd ed. Cambridge: Cambridge University Press.

Birkett, Dea. 2004. "Island of Lost Girls." *New York Times*, October 29. Op-Ed.

Bonanno, Alessandro, and Douglas H. Constance. 1996. *Caught in the Net: The Global Tuna Industry, Environmentalism, and the State.* Lawrence: University Press of Kansas.

Bookchin, Murray. 1982. *The Ecology of Freedom: The Emergence and Dissolution of Hierarchy.* Palo Alto, CA: Cheshire Books.

Borgese, Elisabeth Mann. 1998. *The Oceanic Circle: Governing the Seas as a Global Resource.* New York: United Nations University Press.

Boulding, Elise. 2000. *Cultures of Peace: The Hidden Side of History.* Syracuse: Syracuse University Press.

Breslin, Shaun, and Richard Higgot. 2000. "Studying Regions: Learning from the Old, Constructing the New." *New Political Economy* 5:3, 333–53.

Broecker, Wallace. 1997. "Thermohaline Circulation, the Achilles Heel of Our Climate System: Will Man-made CO_2 Upset the Current Balance?" *Science* 278 (5343): 1582–89.
Brohman, John. 1995. "Universalism, Eurocentrism, and the Ideological Bias in Development Studies: From Modernisation to Neoliberalism." *Third World Quarterly* 16 (1): 121–40.
Brown, Davic, and Robert Pomeroy. 1999. "Co-Management of Caribbean Community (CARICOM) Fisheries." *Marine Policy* 23 (6): 549–70.
Bruinsma, Jelle. 2003. *World Agriculture: Towards 2015:2030, an FAO Perspective*. London: Earthscan.
Bryant, Raymond. 1998. "Resource Politics in Colonial South-East Asia: A Conceptual Analysis." In *Environmental Challenges in South-East Asia*, ed. Victor King, 29–51. Surrey, UK: Curzon Press/Nordic Institute of Asian Studies.
Buck, Susan. 1998. "No Tragedy of the Commons." In *Green Planet Blues: Environmental Politics from Stockholm to Kyoto*, ed. K. Conca and G. Dabelko, 48–54. 3rd ed. Boulder, CO: Westview Press.
Bunker, Stephen. 1985. *Underdeveloping the Amazon: Extraction, Unequal Exchange, and the Failure of the Modern State*. Chicago: University of Chicago Press.
Burke, Lauretta, et al. 2001. *Pilot Analysis of Global Ecosystems: Coastal Ecosystems*. Washington, DC: World Resources Institute. pdf.wri.org/page_coastal.pdf (accessed June 14, 2004).
Burke, Lauretta, Liz Selig, and Marc Spalding. 2002. *Reefs at Risk in Southeast Asia*. Washington, DC: World Resources Institute, at pdf.wri.org/rrseasia (accessed June 14, 2004).
Burtless, Gary, et al. 1998. *Globaphobia: Confronting Fears about Open Trade*. Washington, DC: Brookings Institution.
Cabioch, Guy, et al. 2003. "Continuous Reef Growth During the Last 23_cal_kyr BP in a Tectonically Active Zone (Vanuatu, Southwest Pacific)." *Quarternary Science Reviews* 22:15–17, 1771–86.
Capra, Fritjof. 2002. *Hidden Connections: Integrating the Biological, Cognitive and Social Dimensions of Life into a Science of Sustainability*. New York: Doubleday.
CARICOM. Caribbean Community Secretariat. www.CARICOM.org (accessed February 17, 2004).
Carricart-Garnet, Juan. 2004. "Sea Surface Temperature and the Growth of West Atlantic Reef Building Coral *Montastraea Annularis*." *Journal of Experimental Marine Biology and Ecology* 302:249–60.
Castells, Manuel. 2001. "The Rise of the Fourth World." In D. Held and A. McGrew, *The Global Transformation Reader: An Introduction to the Globalization Debate*. Cambridge: Polity Press.
Cell, John. 2004. "Colonialism and Colonies." *Encarta Online Encyclopedia*, at encarta.msn.com (accessed November 10, 2004).

Chanjindamanee, Siriporn. 2000. "TUF Makes a Big Dent in Tuna Market." *The Nation—Thailand*. December 27.
Chew, Sing. 2001. *World Ecological Degradation: Accumulation, Urbanization, and Deforestation*. Lanham, MD: AltaMira Press/Rowman & Littlefield.
———. 2002. "Globalization, Ecological Crisis, and Dark Ages." *Global Society* 16 (4): 333–56.
Christensen, Villy, Silvie Guenette, Johanna Heymans, Carl Walters, Reginald Watson, Dirk Zeller, and Daniel Pauly. 2003. "Hundred Year Decline in North Atlantic Predatory Fishes." *Fish and Fisheries* 4 (1): 1–24.
Chua, Amy. 2003. *World on Fire: How Exporting Free Market Democracy Breeds Ethnic Hatred and Global Instability*. New York: Doubleday.
Clark, Malcolm. 1999. "Fisheries for Orange Roughy on Seamounts in New Zealand." *Oceanologica Acta* 22 (6): 593–602.
Cole, Hannah. 2003. "Contemporary Challenges: Globalisation, Global Interconnectedness and That 'There Are Not Plenty more Fish in the Sea': Fisheries, Governance and Globalisation: Is There a Relationship?" *Ocean and Coastal Management* 46:77–102.
Coleman, Felicia, et al. 2004. "The Impact of United States Recreational Fisheries on Marine Fish Populations." *Science* 305 (5692): 1958–59.
Conca, Ken. 2001. "Consumption and Environment in a Global Economy." *Global Environmental Politics* 1 (3): 11–30.
Conover, David, and Stephen Munch. 2002. "Sustaining Fisheries Yields Over Evolutionary Time Scales." *Science* 297 (5578): 94–97.
Costanza, Robert. 2000. "The Ecological, Economic, and Social Importance of the Oceans." In *Seas at the Millenium: An Environmental Evaluation*, ed. Charles Sheppard. New York: Pergamon Press.
Coulter, Daniel. 2002. "Globalization of Maritime Commerce: The Rise of the Hub Ports." In *Globalization and Maritime Power*, ed. Sam Tangredi. Washington, DC: Institute for National Strategic Studies, National Defense University.
Cunningham, Richard. 1997. "The Biological Impacts of 1492." In *The Indigenous People of the Caribbean*, ed. S. Wilson. Tallahassee: University Press of Florida.
Daly, Herman, and John Cobb. 1989. *For the Common Good: Redirecting Economy Toward Community, the Environment, and a Sustainable Future*. Boston: Beacon Press.
Day, Trevor. 1999. *Ecosystem: Oceans*. New York: Facts on File, Inc.
Desch, Michael, Jorge Domínguez, and Andrés Serbin, eds. 1998. *From Pirates to Drug Lords: The Post-Cold War Caribbean Security Environment*. Albany: State University of New York Press.
DeWalt, Billie, Philippe Vergne, and Mark Hardin. 1996. "Shrimp Aquaculture Development and the Environment: People, Mangroves and Fisheries on the Gulf of Fonseca, Honduras." *World Development* 24 (7): 1193–1208.
Dobson, Andrew. 2003. *Citizenship and the Environment*. New York: Oxford University Press.
Douglas, Bruce, Michael Kearney, and Stephan Leatherman. 2001. *Sea Level Rise: History and Consequence*. New York: Academic Press.

Douglas, Mike. 2002. "Globalization, Intercity Competition and the Rise of Civil Society: Towards Livable Cities in Pacific Asia." *Asian Journal of Social Science* 30 (1): 129–49.

Doulman, David. 1993. "Community-Based Fishery Management: Towards the Restoration of Traditional Practices in the South Pacific." *Marine Policy* 17 (2): 108–17.

Dragsbaek Schmidt, Johannes. 2000. "Neoliberal Globalization, Social Welfare and Trade Unions in Southeast Asia." In *Globalization and the Politics of Resistance*, ed. Barry Gills, 220–40. New York: St. Martin's Press.

Dujan, Valerie. 2002. "Local Actors, Nation States, and Their Global Environment: Conceptualizing Successful Resistance to Anti-Social Impacts of Globalization." *Critical Sociology* 28 (3): 371–88.

Dunlap, R., G. Gallup, and A. Gallup. 1993. "Health of the Planet: Results of a 1992 International Environmental Opinion Survey of Citizens in 24 Nations." Princeton, NJ: The George H. Gallup International Institute.

Dupuy, Alex. 2001. "The New World Order, Globalization, and Caribbean Politics." In *New Caribbean Thought: A Reader*, ed. B. Meeks and F. Lindahl. Kingston, Jamaica: University of the West Indies.

Dyurgerov, Mark. 2003. "Mountain and Subpolar Glaciers Show an Increase in Sensitivity to Climate Warming and Intensification of the Water Cycle." *Journal of Hydrology*, in press.

Eades, J. S. 1999. "High-Speed Growth, Politics and the Environment in East and Southeast Asia." In *Integrated Environmental Management: Development, Information and Education in the Asian-Pacific Region*, ed. Yasumasa Itakura, et al., 1–18. Boca Raton, FL: Lewis Publishers.

Earle, Sylvia. 1995. *Sea Change: A Message of the Oceans*. New York: Fawcett Columbine.

Eckersely, Robyn. 1992. *Environmentalism and Political Theory: Towards an Ecocentric Approach*. Albany: State University of New York Press.

Economic Commission on Latin America and the Caribbean. 2003. "Social Panorama of Latin America, 2002–2003," LC/G.2218. United Nations Publications.

Economist. 2001. "Radicalism, Asian Style" 358:78, 8214.

Edinger, Evan, et al. 1998. "Reef Degradation and Coral Biodiversity in Indonesia: Effects of Land Based Pollution, Destructive Fishing Practices and Changes Over Time." *Marine Pollution Bulletin* 36 (8): 617–30.

Elu, Juliet. 2000. "The Journey So Far: The Effect of Structural Adjustment Programme (SAP), Sustainable Growth, and Development in the Caribbean Region." *The Western Journal of Black Studies* 24:4, via Infotrac.

Emanuel, Kerry. 2005. "Increasing Destructiveness of Tropical Cyclones over the Past 30 Years." *Nature* 436: 7051.

Enloe, Cynthia. 1990. *Bananas, Beaches, and Bases: Making Feminist Sense of International Politics*. Berkeley: University of California Press.

Eriksson, Mikael, Peter Wallensteen, and Margareta Sollenberg. 2003. "Armed Conflicts 1989–2002." *Journal of Peace Research* 40 (5): 593–607.

Escobar, Arturo. 1995. *Encountering Development: The Making and Unmaking of the Third World.* Princeton: Princeton University Press.

———. 2003. "Displacement, Development, and Modernity in the Columbian Pacific." *International Social Science Journal* 175: 157–67.

———. 2004. "Beyond the Third World: Imperial Globality, Global Coloniality, and Anti-Globalization Movements." *Third World Quarterly* 25 (1): 207–30.

Evans, Peter. 2004. "Development as Institutional Change: The Pitfalls of Monocropping and the Potentials of Deliberation." *Studies in Comparative International Development* 38 (4): 30–52.

Feely, Richard, et al. 2004. "Impact of Anthropocentric CO_2 on the $CaCO_3$ System in the Oceans." *Science* 305 (5682): 362–66.

Food and Agricultural Organization (FAO). 1997. "Western Central Pacific." In *Review of the State of the World Fishery Resources: Marine Fisheries* (Circular 920 FIRM/C920). Rome: FAO.

———. 1999. "Technical Paper 386." In *Managing Fishing Capacity: Selected Papers on Underlying Concepts and Issues*, at www.fao.org/DOCREP/003/X2250E/x2250e00.htm#Contents (accessed October 29, 2004).

———. 2002. *State of the World Fisheries and Aquaculture*, at www.fao.org/sof/sofia/index_en.htm (accessed February 24, 2003).

———. 2003a. *FAO Yearbook Fishery Statistics: Capture Production.* Vol. I. Rome: FAO.

———. 2003b. *World Agriculture: Towards 2015/2030, an FAO Perspective.* Available online at www.fao.org/DOCREP/005/y4252E/Y4252Eoo.htm.

Friedman, Milton. 1962. *Capitalism and Freedom.* Chicago: University of Chicago Press.

Fritsch, Peter. 2004. "As Shrimp Industry Thrives in Vietnam, Trade Fight Looms." *Wall Street Journal.* October 21.

Fry, Greg. 1997. "Framing the Islands: Knowledge and Power in Changing Australian Images of 'The South Pacific.'" *Contemporary Pacific* 9 (2): 305–44.

Fukuyama, Francis. 1992. *The End of History and the Last Man.* New York: Free Press.

Gadamer, Hans-Georg. 1993. *Truth and Method.* 2nd revised edition. New York: Continuum Press.

Gage, John. 2004. "Diversity in Deep-sea Benthic Macrofauna: The Importance of Local Ecology, the Larger Scale, History and the Antarctic." In *Deep Sea Research II* 51: 14–16, 1689–1708.

Garcia, S. M., and C. Newton. 1997. "Current Situation, Trends, and Prospects in World Capture Fisheries." In *Global Trends: Fisheries Management*, ed. Ellen Pikitch, et al. Bethesda, MD: American Fisheries Society.

Gardner, Toby, et al. 2003. "Long-term Region-wide Declines in Caribbean Corals." *Science* 301 (5635): 958–61.

Geider, Richard. 2001. "Primary Productivity of Planet Earth: Biological Determents and Physical Constraints on Terrestrial and Aquatic Habitats." *Global Change Biology* 7: 849–82.

Gell-Mann, Murray. 1994. *The Quark and the Jaguar: Adventures in the Simple and Complex.* Boston: Little, Brown.

Georgescu-Roegen, Nicholas. 1971. *The Entropy Law and the Economic Process.* Cambridge, MA: Harvard University Press.

Giddens, Anthony. 1979. *Central Problems in Social Theory: Action, Structure, and Contradiction in Social Analysis.* Berkeley: University of California Press.

Gillett, R. D. 2002. *Pacific Island Fisheries: Regional and Country Information. Asia.* Pacific Fishery Commission, FAO Regional Office for Asia and the Pacific. Bangkok: RAP Publication 2002/13.

Goodland, Robert. 1995. "The Concept of Sustainability." *Annual Review of Ecology and Systematics* 26: 1–24.

Goss, Jasper, David Burch, and Roy Rickson. 2000. "Agri-food Restructuring and the Third World Transnationals: Thailand, the CP Group and the Global Shrimp Industry." *World Development* 28 (3): 513–30.

Grassl, Hartmut. 2001. "Climate and Oceans." In *Ocean Circulation and Climate: Observing and Modeling the Global Ocean,* ed. Gerold Siedler, John Church, and John Gould. New York: Academic Press.

Griffith, Ivelaw. 1998. "The Geography of Drug Trafficking in the Caribbean." In *From Pirates to Drug Lords: The Post–Cold War Caribbean Security Environment,* ed. M. Desch, J. Domínguez, and A. Serbin. Albany: State University of New York Press.

Grubuhel, Clemons, et al. 2003. "Socioeconomic Metabolism and Colonization of Natural Processes in SangSaeng Village: Material and Energy Flows, Land Use, and Cultural Change in Northeast Thailand." *Human Ecology: An Interdisciplinary Journal,* 31 (1): 53–87.

Guttal, Shalmali. 2001. "The Emperor's New Clothes: The Asian Development Bank's Poverty Reduction Program." In *Profiting from Poverty: The ADB, Private Sector, and Development in Asia,* 3–10. Bangkok: Focus on the Global South, at www.focusweb.org (accessed June 13, 2004).

Hadad, Nadia. 2001. "ADB in Indonesia: Alleviate Poverty or Enhance Poverty?" In *Profiting from Poverty: The ADB, Private Sector, and Development in Asia,* 47–48. Bangkok: Focus on the Global South, at www.focusweb.org (accessed June 13, 2004).

Hardin, Garrett. 1968. "The Tragedy of the Commons." *Science* 162:1243–48.

Hardoy, Jorge, Diana Mitlin, and David Satterthwaite. 2001. *Environmental Problems in an Urbanizing World.* London: Earthscan.

Harris, Paul. 2003. *Global Warming and East Asia: The Domestic and International Politics of Climate Change.* New York: Routledge.

Harrison, Neil. 2000. *Constructing Sustainable Development.* Albany: State University of New York Press.

Hau'ofa, Epeli. 1994. "Our Sea of Islands." *Contemporary Pacific* 6 (1): 148–61.

Hawkins, Julie, and Calumm Roberts. 2004. "Effects of Fishing on Sex-changing Caribbean Parrotfishes." *Biological Conservation* 115 (2): 213–26.

Hawkins, Ronnie. 2002. "Seeing Ourselves as Primates." *Ethics and the Environment* 7 (2): 60–103.
Hay Brown, Matthew. 2004. "Many Haitians Can't Find Food." *Orlando Sentinel*, A1, A18. March 7.
Hayles, N. Katherine. 1995. "Searching for Common Ground." In *Reinventing Nature: Responses to Postmodern Deconstruction*, ed. M. Soulé and G. Lease. Washington, DC: Island Press.
Hedman, Eva-Lottta. 2001. "Contesting State and Civil Society: Southeast Asian Trajectories." *Modern Asian Studies* 35 (4): 921–51.
Held, David, et al. 1999. *Global Transformations: Politics, Economics and Culture*. Stanford, CA: Stanford University Press.
Henry, Paget. 1998. "Philosophy and the Caribbean Tradition." *Small Axe* 4:3–28.
Hettne, Björn. 1999. "Prologue to the Five Volumes." In *Globalism and the New Regionalism*, ed. Björn Hettne, András Inotai, and Osvaldo Sunkel. New York: St. Martin's Press.
Hettne, Björn, András Inotai, and Osvaldo Sunkel, eds. 2001. *Comparing Regionalisms: Implications for Global Development*. New York: Palgrave.
Holmlund, Cecilia, and Monica Hammer. 1999. "Ecosystem Services Generated by Fish Populations." *Ecological Economics* 29 (2): 253–68.
Hornberg, Alf. 2001. "Vital Signs: An Ecosemiotic Perspective on the Human Ecology of Amazonia." *Sign System Studies* 29 (1): 121–52.
Huang, Chi-Yue, et al. 1997. "Surface Ocean and Monsoon Climate Variability in the South China Sea since the Last Glaciation." *Marine Micropaleontology* 32 (1–2): 71–94.
Hughes, T. P., et al. 2003. "Climate Change, Human Impacts, and the Resilience of Coral Reefs." *Science* 301: 929–33.
Hunt, Colin. "Economic Globalisation Impacts on Pacific Island Environments and Aid Implications." In *Economic Globalisation: Social Conflict, Labour and Environmental Issues*, ed. C. Tisdell and R. K. Sen. Cheltenham, UK: Edward Elgar, 2003.
Hunter Wade, Robert. 2004. "Is Globalization Reducing Poverty and Inequality?" *World Development* 32 (4): 567–89.
International Monetary Fund (IMF). 2004. "World Economic Outlook: The Global Demographic Transition," at www.imf.org/external/pubs/ft/weo/2004/02/index.htm (accessed October 27, 2004).
International Panel on Climate Change (IPCC). 2001. *Climate Change 2001: Synthesis Report*. Cambridge: Cambridge University Press.
Jackson, Jeremy B. C., et al. 2001. "Historical Overfishing and the Recent Collapse of Coastal Ecosystems." *Science* 293 (5530): 629–39.
Jackson, Moana. 1993. "Indigenous Law and the Sea." In *Freedom for the Seas in the 21st Century: Ocean Governance and Environmental Harmony*, ed. J. Van Dyke, et al. Washington, DC: Island Press.
Jacques, Peter. 2001. "Teaching Peacefully: The Ocean Policy as a Case." *The International Journal of Humanities and Peace* 17:1.

———. 2003. "A Green Peace? Connections between Environmental Policy and Foreign Policy." PhD diss., Northern Arizona University.

Jacques, Peter, and Zachary Smith. 2003. *Ocean Politics and Policy: A Handbook*. Santa Barbara, CA: ABC/Clio.

Jiang, Yihang, et al. 2001. "Megacity Developments: Managing Impacts on the Marine Environment." *Ocean and Coastal Development* 44:293–318.

Johannes, Robert. 1978. "Traditional Marine Conservation Methods in Oceania and their Demise." *Annual Review of Ecology and Systematics* 9:349–64.

Johnson, Gregory, and Alejandro Orsi. 1997. "Southwest Pacific Ocean Water-mass Changes between 1968/69 and 1990/91." In *Journal of Climate* 10 (February): 306–16.

Joseph, Garnette. 1997. "Five Hundred Years of Indigenous Resistance." In *The Indigenous People of the Caribbean*, ed. S. Wilson. Gainesville: University Press of Florida.

Juda, Lawrence. 1996. *International Law and Ocean Use Management: The Evolution of Ocean Governance*. New York: Routledge.

Keohane, Robert, and Elinor Ostrom, eds. 1995. *Local Commons and Global Interdependence: Heterogeneity and Cooperation in Two Domains*. London: Sage Publications.

Kim, Sangmoon, and Eui-Hang Shin. 2002. "Attitudinal Analysis of Globalization and Regionalization in International Trade: A Social Network Approach." *Social Forces* 81 (2): 445–68.

Kleypas, Joan, et al. 1999. "Geochemical Consequences of Increased Atmospheric Carbon Dioxide on Coral Reefs." *Science* 284 (5411): 125–29.

Knight, W. Andy, and Randolph Persaud. 2001. "Subsidiarity Regional Governance, and Caribbean Security." *Latin American Politics and Society* 43:1. Also available online via Infotrac.

Knippers Black, Jan. 1999. *Inequity in the Global Village: Recycled Rhetoric and Disposable People*. West Hartford: Kumarian Press.

Knowlton, Nancy. 2001. "Coral Reef Biodiversity—Habitat Size Matters." *Science* 292 (5521): 1493–95.

Krabil, W. W., et al. 2000. "Greenland Ice Sheet." *Science* 289 (5478): 428–31.

Kütting, Gabriela. 2004. *Globalization and the Environment: Greening Political Economy*. Albany: State University of New York Press.

Lang, Chris. 2001. "Shrimps, Mangroves, and the World Bank." World Rainforest Movement, at www.wrm.org.uy/bulletin/51/Vietnam.html (accessed June 19, 2004).

Larkin, P. A. 1978. "Fisheries Management—an Essay for Ecologists." *Annual Review of Ecology and Systematics* 9:57–73.

Lee, Josephine. 2003. "Fowl Play." *Forbes* 171 (6): 182–84.

Lee, Kai. 1993. *Compass and Gyroscope: Integrating Science and Politics for the Environment*. Washington, DC: Island Press.

Lehody, Patrick, Fei Chai, and John Hampton. 2003. "Modelling Climate Variability of Tuna Populations from a Coupled Ocean-Biogeochemical-Populations Dynamics Model." *Fisheries Oceanography* 12 (4/5): 483–94.

Lemay, Michele. 1998. "Coastal and Marine Resources Management in Latin America and the Caribbean." Inter-American Development Bank, at www.iadb.org/sds/doc/1097eng.pdf (accessed July 29, 2004).

Leopold, Aldo. [1949] 1966. *A Sand County Almanac, with Essays on Conservation from Round River.* Cambridge: Cambridge University Press.

Lepowski, Maria. 1994. "Women, Men, and Aggression in an Egalitarian Society." *Sex Roles* 30 (3/4): 199, Infotrac Online.

Lewsey, Clement, Gonzolo Cid, and Edward Kruse. 2004. "Assessing Climate Change Impacts on Coastal Infrastructure in the Eastern Caribbean." *Marine Policy* 28 (5): 393–409.

Linton, Dulcie, and George Warner. 2003. "Biological Indicators in the Caribbean Coastal Zone and their Role in Integrated Coastal Management." *Ocean and Coastal Management* 46: 261–76.

Lobban, Christopher, and Maria Schefter. 1997. *Tropical Pacific Island Environments.* Mangilao: University of Guam Press.

Lomborg, Bjørn. 2001. *The Skeptical Environmentalist: Measuring the Real State of the World.* New York: Cambridge University Press.

Longhurst, Alan. 2001. *Ecological Geography of the Sea.* New York: Academic Press.

Lu, F. 1998. *Output Data on Animal Products in China: How Much They Are Overstated.* Beijing: China Center for Economic Research.

Lubchenko, Jane. 2003. "Waves of the Future: Sea Changes for a Sustainable World." In *Worlds Apart: Globalization and the Environment,* ed. James Gustave Speth. Washington, DC: Island Press.

Luxner, Larry. 1999. "CARICOM: 25 Years of a United Caribbean Voice." *Americas* (English Edition) 51:1, 56.

Macleod, Donald. 2004. "Selling Space: Power and Resource Allocation in a Caribbean Coastal Community." In *Confronting Environments: Local Understanding in a Globalizing World,* ed. James Carrier. Lanham, MD: AltaMira Press/Rowman & Littlefield.

Marcuse, Herbert. 1964. *One Dimensional Man.* Boston: Beacon Press.

Marte, Clarissa. 2003. "Larviculture of Marine Species in Southeast Asia: Current Research and Industry Prospects." *Aquaculture* 227 (1–4): 293–304.

Martin-Ramos, Jesus. 2003. "Empiricism in Ecological Economics: A Perspective from Complex Systems Theory." *Ecological Economics* 46 (3): 387–99.

Martinez-Alier, Juan. 1995. "Distributional Issues in Ecological Economics." *Review of Social Economy* 54 (4): 511–29.

———. 2000. "Ecological Conflicts and Valuation—Mangroves vs. Shrimp in the Late 1990s," at www.seatrans.net/information/tools/multicriteria analysis (accessed June 2, 2004).

Mason, B. J. 1993. "The Role of the Oceans in Climate Change." *Contemporary Physics* 34 (1): 19–30.

Matthews, Richard, and Ted Gaulin. 2001. "Conflict or Cooperation? The Social and Political Impacts of Resource Scarcity on Small Island States." *Global Environmental Politics* 1 (2): 48–70.

McCook, Stuart. 2002. *States of Nature: Science, Agriculture, and Environment in the Spanish Caribbean.* Austin: University of Texas Press.

McCulloch, Malcolm, et al. 2003. "Coral Record of Increased Sediment Flux to the Inner Great Barrier Reef Since European Settlement." *Nature* 421:727–30.

McGoodwin, James. 1990. *Crisis in the World's Fisheries: People, Problems, and Policies.* Stanford, CA: Stanford University Press.

McMahon, Robert. 1999. *The Limits of Empire: The United States and Southeast Asia Since World War II.* New York: Columbia University Press.

Meeks, Brian. 2001. "On the Bump of a Revival." In *New Caribbean Thought: A Reader*, ed. B. Meeks and F. Lindahl. Kingston, Jamaica: University of the West Indies.

Meeks, Brian, and Folke Lindahl, eds. 2001. *New Caribbean Thought: A Reader.* Kingston, Jamaica: University of the West Indies.

Mendoza, Jeremy, and Asdrubal Larez. 2004. "A Biomass Dynamics Assessment of the Southern Caribbean Snapper-grouper Fishery." *Fishery Research* 66 (2–3): 129–44.

Merchant, Carolyn. 1980. *The Death of Nature: Women, Ecology, and the Scientific Revolution.* San Francisco: Harper and Row.

Milich, Lenard. 1999. "Resource Management versus Sustainable Livelihoods: The Collapse of the Newfoundland Cod Fishery." *Society and Natural Resources* 12:625–42.

Moberg, Frederik, and Carl Folke. 1999. "Ecological Goods and Services of Coral Reef Ecosystems." *Ecological Economics* 29 (2): 215–33.

Mol, Arthur. 2001. *Globalization and Environmental Reform: The Ecological Modernization of the Global Economy.* Cambridge, MA: MIT Press.

Mörner, Nils-Axel, Michael Tooley, and Göran Possnert. 2003. "New Perspectives for the Future of the Maldives." *Global and Planetary Change.*

Mulekom, Leo van, et al. 2003. "Trade and Export Orientation of Fisheries in Southeast Asia: Under-priced Exports at the Expense of Domestic Food Security and Local Economies." Partnerships for Environmental Management for the Seas of East Asia, at www.pemsea.org/downloads_pdf/abstracts/A3/3LeoMulekom-A-3_Trade%20and%20Export.pdf (accessed June 19, 2004).

Muradian, Rolden, and Juan Martinez-Alier. 2001a. "Trade and the Environment: From a 'Southern' Perspective." *Ecological Economics* 36:281–97.

———. 2001b. "South-North Materials Flow: History and Environmental Repercussions." *Innovation: The European Journal of Social Sciences* 14 (2): 171–89.

Narine, Shaune. 2002. *Explaining ASEAN: Regionalism in Southeast Asia.* Boulder, CO: Lynne Reiner Publishers.

Narsalay, Raghav. 2001. "South Asia Growth Quadrangle: Some Developmental and Political Contradictions." In *Profiting from Poverty: The ADB, Private Sector, and Development in Asia*, 34–40. Bangkok: Focus on the Global South, at www.focusweb.org (accessed June 13, 2004).

Nash, Roderick. 1990. *American Environmentalism: Readings in Conservation History*. New York: Knopf.

Nerem, R. S., and G. T. Mitchum. 2000. "Observation of Sea Level Change from Satellite Altimetry." In *Sea Level Rise: History and Consequences*, ed. Douglas, et al. New York: Academic Press.

Norris, Scott, et al. 2002. "Thinking like an Ocean: Ecological Lessons from Marine Bycatch." *Conservation in Practice* 3:4, at www.conbio.org/inpractice/article34THI.cfm (accessed November 1, 2004).

Nöth, Winfried. 2001. "Ecosemiotics and the Semiotics of Nature." *Sign System Studies* 29 (1): 71–81.

Nunn, Patrick, et al. 2002. "Late Quarternary Sea Level and Tectonic Changes in Northeast Fiji." *Marine Geology* 187 (3–4): 299–311.

Oceana. 2004. "Oceana Victory: Royal Caribbean to Clean up Its Act." Press release. May 3, at www.stopcruisepollution.com (accessed June 20, 2004).

Oeschlaeger, Max. 2001. "Ecosemiotics and the Sustainability Transition." *Sign System Studies* 29 (1): 219–36.

Okey, Thomas. 2003. "Membership of the Eight Regional Fishery Management Councils in the United States: Are Special Interests Over-represented?" *Marine Policy* 27 (3): 193–206.

Oliver, José. 1997. "The Taino Cosmos." In *The Indigenous People of the Caribbean*, ed. S. Wilson, 140–53. Tallahassee: University Press of Florida.

Oneal, John, and Bruce Russett. 1997. "The Classical Liberals were Right: Democracy, Interdependence, and Conflict, 1950–1980." *International Studies Quarterly*, 41 (2): 267–94.

Organisation of Eastern Caribbean States (OECS). 2004. "About the OECS Secretariat," at www.oecs.org/about.htm (accessed February 27, 2004).

Ostergren, David, and Peter Jacques. 2002. "A Political Economy of Russian Nature Conservation Policy: Why Scientists Have Taken a Back Seat." *Global Environmental Politics* 2 (4): 102–24.

Ostrom, Elinor, and Christopher Field. 1999. "Revisiting the Commons: Local Lessons, Global Changes." *Science* 284 (5412): 278–83.

Pacific Islands Forum. 2003 Forum Communiqué, Auckland, Australia, at www.forumsec.org.fj/docs/Communique/2003%20Communique.pdf (accessed January 23, 2004).

Padma, T. V. 2004. "Mangroves Can Reduce the Impact of Tsunamis." December 30, at www.scidev.net (accessed February 15, 2005).

Paehlke, Robert. 2004. *Democracy's Dilemma: Environment, Social Equity, and the Global Economy*. Cambridge, MA: MIT Press.

Pauly, Daniel, et al. 2002. "Toward Sustainability in World Fisheries." *Nature* 418 (6898): 689–95.

Paramaswaran, K., Sandya Nair, and K. Rajeev. 2004. "Impact of Indonesian Forest Fires

During the 1997 El Niño on the Aerosol Distribution over the Indian Ocean." *Advances in Space Research* 33 (7): 1098–1104.

Potash Corporation. 2004. "News Releases." May 10, at www.potashcorp.com/investor_relations/news_releases/index.zsp?newsid=174 904 (accessed June 19, 2004).

Prager, Ellen, and Sylvia Earle. 2000. *The Oceans*. New York: McGraw Hill.

Princen, Thomas. 2003. "Principles for Sustainability: From Cooperation and Efficiency to Sufficiency." *Global Environmental Politics* 3 (1): 33–50.

Radford Ruether, Rosemary. 1994. "Tourists Must Look beyond the Walls to See the Scars." *National Catholic Reporter*, 30:12, 28. January 21.

Ram-Bidesi, Vina, and Martin Tsamenyi. 2004. "Implications of the Tuna Management Regime for Domestic Industry Development in the Pacific Island States." In *Marine Policy* 28 (5): 383–92.

Randall, Stephen, and Graeme Mount. 1998. *The Caribbean Basin: An International History*. New York: Routledge.

Reid, Christopher, et al. 2003. "An Analysis of Fishing Capacity in the Western and Central Pacific Ocean Tuna Fishery and Management Implications." *Marine Policy* 27 (6): 449–69.

Renard, Yves, Allan Smith, and Vijay Krishnarayan. 2000. "Do Reefs Matter? Coral Reef Conservation, Sustainable Livelihoods, and Poverty Reduction." Paper presented at the Managing Space for Sustainable Living in Small Island Developing States for the Caribbean Natural Resources Institute, at www.canari.org/274reefs.pdf (accessed February 27, 2004).

Revkin, Andrew. 2002. "Can Global Warming Be Studied Too Much?" *New York Times*, D1. December 3.

———. 2003. "Huge Ice Shelf Is Reported to Break up in Canada." *New York Times*, A10. September 23.

———. 2004. "Carbon Dioxide Extends Its Harmful Reach to Oceans." *New York Times*. July 20.

Ridgeway, Sharon. 1996. "Critique of the Mechanistic Paradigm for Environmental Policy." PhD diss., Northern Arizona University.

Rietkerk, Max, et al. 2004. "Self-organized Patchiness and Catastrophic Shifts in Ecosystems." *Science* 305 (5692): 1926–29.

Rigby, Andrew. 1997. "'Gram Swaraj'" versus 'Globalization.'" *Peace and Change* 22 (4): 381–414.

Rihani, Samir. 2002. *Complex Systems Theory and Development Practice: Understanding Non-Linear Realities*. New York: Zed Books.

Robertson, Robbie. 2003. *The Three Waves of Globalization: A History of a Developing Global Consciousness*. New York: Zed Books.

Rogers, Katrina. 1996. *Toward a Postpositivist World: Hermeneutics for Understanding International Relations, Environment, and Other Important Issues of the Twenty-first Century*. New York: Peter Lang.

———. 1997. "Ecological Security and Multinational Corporations." In *Environmental Change and Security Project Report 3*, Woodrow Wilson Center. Spring.
Rolfe, Jim. 2001. "Peacekeeping the Pacific Way in Bougainville." *International Peacekeeping* 8(4): 38–56.
Rosenau, James. 1990. *Turbulence in World Politics: A Theory of Change and Continuity.* Princeton, NJ: Princeton University Press.
Roughan, John. 1997. "Solomon Islands Nongovernmental Organizations: Major Environmental Actors." *Contemporary Pacific* 9 (1): 157–67.
Roy, Arundhati. 2002. "Shall We Leave It to the Experts?" *The Nation* 274 (6): 16–20, Ebscohost online.
Sabia, Dan. 2003. "Immanent Critique and Utopianism." Paper presented at the annual meeting of the Society for Utopian Studies, San Diego. October 31.
Sabine, Christopher, et al. 2004. "The Oceanic Sink for Anthropocentric CO_2." *Science* 305 (5682): 367–71.
Safina, Carl. 1998. *Song for a Blue Ocean: Encounters along the World's Coasts and beneath the Seas.* New York: Henry Holt.
Samou, Saolme. 1999. "Marine Resources." In *Strategies for Sustainable Development: Experiences from the Pacific*, ed. John Overton and Regina Scheyvens, 142–54. New York: Zed Books.
Satria, Arif, and Yoshiaki Matsuda. 2004. "Decentralization of Fishery Management in Indonesia." *Marine Policy* 28 (5): 437–50.
Scheyvens, Regina. 1999. "Culture and Society." In *Strategies for Sustainable Development: Experiences from the Pacific*, ed. John Overton and Regina Scheyvens, 48–63. New York: Zed Books.
Scheyvens, Regina, and Leonard Lagisa. 1998. "Women, Disempowerment, and Resistance: An Analysis of Logging and Mining Activities in the Pacific." *Singapore Journal of Tropical Geography* 19 (1): 51–70.
Scheyvens, Regina, and Ross Cassells. 1999. "Logging in Melanesia." In *Strategies for Sustainable Development: Experiences from the Pacific*, ed. John Overton and Regina Schnaiberg, Allan. 1980. *The Environment: From Surplus to Scarcity.* New York: Oxford University Press.
Schulkin, Andrew. 2002. "Safe Harbors: Crafting an International Solution to Cruise Ship Pollution." *Georgetown International Environmental Law Review* 15(1): 105–58.
Schurman, Rachel. 1998. "Tuna Dreams: Resource Nationalism and the Pacific Islands' Tuna Industry." *Development and Change* 29 (1): 107–36.
Sea Around Us Project. 2004. "Large Marine Ecosystems," at www.seaaroundus.org (accessed July 21, 2004).
Searle, John. 1995. *The Construction of Social Reality.* New York: Free Press.
Secretariat of the Pacific Community. 2003. Official website, at www.spc.org.nc/ (accessed January 23, 2004).
Seuront, Laurent, and Nicolas Spilmont. 2002. "Self-organized Criticality in Intertidal Microphytobenthos Patch Patterns." *Physica* 313 (3–4): 513–39.

Sharp, Basil. 2005. "ITQs and beyond in New Zealand Fisheries." In *Evolving Property Rights in Marine Fisheries*, ed. D. Leal. Lanham, MD: Rowman & Littlefield.

Sheller, Mimi. 2003. *Consuming the Caribbean: From Arawaks to Zombies*. New York: Taylor and Francis.

Sheppard, Charles. 2003. "Predicted Recurrences of Mass Coral Mortality in the Indian Ocean." *Nature* 425 (6955) 294–97.

Shotton, R. 1997. "Southwest Pacific." *Review of the State of the World Fisheries*, FAO Circular No. 920 FIRM/C920 (En), at www.fao.org/docrep/003/w4248e/w4248e00.htm (accessed January 28, 2004).

Simpson, Sarah. 2001. "Cyanide Fishing Threatens Many of the Last Pristine Coral Reefs in Southeast Asia. Will an Ambitious Program to Clean up the Marine Aquarium Trade Be Enough to Save Them?" *Scientific American* 285 (1): 82–90.

Singh, Bhawan. 1997. "Climate-related Global Changes in the Southern Caribbean: Trinidad and Tobago." *Global and Planetary Change* 15 (3–4): 93–111.

Singh, Kavaljit. 1999. *The Globalization of Finance: A Citizens Guide*. New York: Zed Books.

Siregar, P. Raja. 2004. "Large-scale Shrimp Farming and Impacts on Women." World Rainforest Movement, at www.wrm.org.uy/deforestation/shrimp/women.rtf (accessed June 19, 2004).

Soulé, Michael, and Gary Lease, eds. 1995. *Reinventing Nature? Responses to Postmodern Deconstruction*. Washington, DC: Island Press.

South Pacific Regional Environmental Programme. 2002. *Report of the Thirteenth SPREP Meeting of Officials and Report of the Environment Ministers' Forum*, at www.sprep.org.ws/att/publication/000135_13_SPREP_Meeting_report_of_Officials_and_Ministers.pdf (accessed January 28, 2004).

South Pacific Regional Environmental Programme. 2003. Draft. *Synopsis of Sustainable Development in PICs: The Pacific Regional Assessment and Position for BPoA + 10*. On file with author.

South Pacific Women's Bureau. 2003. *Pacific Women's Bureau Strategic Plan 2003*, at www.spc.org.nc/Women/About/PWBSP%20III.doc (accessed January 30, 2004).

Spalding, Mark, Corinna Ravilious, and Edmund Green. 2001. *World Atlas of Coral Reefs*. Los Angeles: University of California Press.

Speth, James Gustave, ed. 2003. *Worlds Apart: Globalization and the Environment*. Washington, DC: Island Press.

Sponsel, Leslie. 2000. "Identities, Ecologies, Rights, and Futures: All Endangered." In *Endangered Peoples of Southeast and East Asia: Struggles to Survive and Thrive*, ed. Leslie Sponsel, 1–22. Westport, CT: Greenwood Press.

Stolton, Sue, and Nigel Dudley. 1999. *Partnerships for Protection: New Strategies for Planning and Management for Protected Areas*. London: Earthscan.

Stonich, Susan, and John Bort. 1997. "Globalization of Shrimp Mariculture: The Impact on Social Justice and Environmental Quality in Central America." *Society and Natural Resources* 10 (2): 161–80.

Subhanrao Pednekar, Sunil. 1995. "NGOs and Natural Resource Management in Mainland Southeast Asia." *Thailand Development Research Institute (TDRI) Quarterly Review* 10 (3): 21–27.

Sued-Badillo, Jalil. 1992. "Facing up to Caribbean History." *American Antiquity* 57 (4): 599–607.

Suter, Keith. 1996. "U.S. Signs on at Last." *Bulletin of the Atomic Scientists* 52 (2): 12–14.

Takashi, Taro. 2004. "The Fate of Industrial Carbon Dioxide." *Science* 305 (5682): 352–53.

Taylor, Peter, and Frederick Buttel. 1992. "How Do We Know We Have Global Environmental Problems? Science and the Globalization of Environmental Discourse." *Geoforum* 23 (3): 405–16.

Thompson, Carol. 2000. "Regional Challenges to Globalization: Perspectives from Southern Africa." *New Political Economy* 5 (1): 41–58.

Tickner, J. Ann. 2004. "The Gendered Frontiers of Globalization." *Globalizations* 1 (1): 15–23.

"Trinidad: Women Own Less than 2% of the Land." 1999. *Women's International News Network* 25 (4): 22.

Tsamenyi, Martin. 1999. "The Institutional Framework for Regional Cooperation in Ocean and Coastal Management in the South Pacific." *Ocean and Coastal Management* 42: 465–81.

Tuchman-Mathews, Jessica. 1989. "Redefining Security." *Foreign Affairs* 68: 162–77.

United Nations. 2001. *United Nations Convention on the Law of the Sea*. At www.un.org/Depts/los/index.htm (accessed July 27, 2004).

United Nations Economic and Social Commission for Asia and the Pacific (UNESCAP). 2003. *Bulletin on Asia-Pacific Perspectives 2003–2004*. New York: United Nations. At www.unescap.org/pdd/publications/bulletin03-04 (accessed June 17, 2004).

Vallega, Adalberto, and Stefano Belfiore. 2002. "Ocean and Coastal Zones in Global Programmes and the Ocean 21 Project." In *Marine Issues: From a Scientific, Political and Legal Perspective*, ed. Peter Ehlers, Elisabeth Mann Borgese, and Rudiger Wolfrum. Norwell, MA: Kluwer Law International.

Van Dyke, Jon, Durwood Zaelke, and Grant Hewison, eds. 1993. *Freedom for the Seas in the 21st Century: Ocean Governance and Environmental Harmony*. Washington, DC: Island Press.

Vannuccini, Stefania. 2003. *Overview of Fish Production, Utilization, Consumption and Trade*. Rome: FAO.

Vidal, John. 2003. "Farmer Commits Suicide at Protests." *The Guardian Unlimited*, September 11, at www.guardian.co.uk/wto/article/0,2763,1039709,00.html (accessed October 22, 2004).

Walker, K. J. 2002. "Environmental Policy in Australia." In *Environmental Politics and Policy in Industrialized Countries*, ed. U. Desai. Cambridge, MA: MIT Press.

Wallerstein, Immanuel. 1989. *The Modern World System*. New York: Academic Press.

Waltz, Kenneth. 1959. *Man, the State, and War: A Theoretical Analysis*. New York, Columbia University Press.

Wapner, Paul. 1996. *Environmental Activism and World Civic Politics*. Albany: State University of New York Press.
Warner, Gary. 1997. "Participatory Management, Popular Knowledge, and Community Empowerment: The Case of Sea Urchin Harvesting in the Vieux Fort Area of St. Lucia." *Human Ecology: An Interdisciplinary Journal* 25 (1): 29–47.
Warr, Peter. 2001. "Poverty Reduction and Sectoral Growth: Evidence from Southeast Asia." Proceedings of WIDER Development Conference on Growth and Poverty May 25–26, Helsinki, Finland. At www.wider.unu.edu/conference/conference-2001-1/development-conference-2000-1-papers.htm (accessed December 21, 2005).
Wartho, Richard, and John Overton. 1999. "The Pacific Islands in the World." In *Strategies for Sustainable Development: Experiences from the Pacific*, ed. John Overton and Regina Scheyvens. New York: Zed Books.
Watson, R., and D. Pauly. 2001. "Systematic Distortions in World Fisheries Catch Trends." *Nature* 414: 534–36.
Webster, P. J. et al. 2005. "Changes in Tropical Cyclone Number, Duration, and Intensity in a Warming Environment." *Science* 16 (309): 1844–46.
Western and Central Pacific Fisheries Convention Preparatory Conference. 2004. "Convention on the Conservation and Management of Highly Migratory Fish Stocks in the Western and Central Pacific Ocean," at www.ocean-affairs.com/convention.html (accessed June 26, 2004).
Westing, Arthur, ed. 1986. *Global Resources and International Conflict: Environmental Factors in Strategic Policy and Action*. New York: Oxford University Press.
Wheeler, Raymond Holder. 1936. "Organismic Logic in the History of Science." *Philosophy of Science* 3 (1): 26–61.
Wilkinson, C. 2002. *Status of Coral Reefs of the World: 2002*. Australian Institute of Marine Science, Townsville, 378.
Williams, Nigel. 1998. "Over-fishing Disrupts Entire Ecosystem." *Science* 279 (5352): 809.
Williams, Peter. 2003. "Overview of the Western and Central Pacific Ocean Tuna Fisheries 2002." Standing Committee on Tuna and Billfish, working paper SCTB16, Gen-1. At www.spc.int/OceanFish/Html/SCTB/SCTB16/gen1.pdf (accessed January 7, 2004).
Williamson, John. 1993. "Democracy and the 'Washington Consensus.'" *World Development* 21: 1329–36.
Wilson, Samuel. 1997. "Introduction to the Study of Indigenous People of the Caribbean." In *The Indigenous People of the Caribbean*, ed. S. Wilson, 1–8. Gainesville: University Press of Florida.
Wolff, Edward. 2002. *Top Heavy: The Increasing Inequality of Wealth in America and What Can Be Done about It*. New York: The New Press/W. W. Norton.
Wood, Richard, Anne Keen, John Mitchell, and Jonathan Gregory. 1999. "Changing Spatial Structure of the Thermohaline Circulation in Response to Atmospheric CO_2 Forcing in a Climate Model." *Nature* 399: 572–76.

Wood, Robert. 2000. "Caribbean Cruise Tourism: Globalization at Sea." *Annals of Tourism Research* 27 (2): 345–70.

Wong, Annie, et al. 2001. "Freshwater and Heat Changes in the North and South Pacific Oceans between the 1960s and 1985–94." *Journal of Climate* 14: 1613–33.

World Bank. 1999. "Medium Size Project Brief: Samoa-marine Biodiversity Protection and Management," at www.gefweb.org/wprogram/Jan99/wb/Samoa.doc (accessed January 14, 2004).

———. 2000. *Voices from the Village: A Comparative Study of Coastal Resource Management in the Pacific Islands*. Pacific Islands Discussion Paper Series, #9A, at www.lnweb18.world bank.org/eap/eap.nsf/Attachments/Voices + from + the + village/$File/3_00 + SUMMARY + BOOK + (web).pdf (accessed January 14, 2004).

———. 2002. *Globalization, Growth and Poverty: Building an Inclusive World Economy*. New York: World Bank/Oxford University Press.

World Commission on Environment and Development. 1987. *Our Common Future*. Oxford: Oxford University Press.

World Conservation Union (IUCN). 2004. *2004 IUCN Red List of Threatened Species*, at www.iucnredlist.org accessed (accessed November 18, 2004).

World Resource Institute. 2003a. "Income, Distribution and Poverty." www.earthtrends.wri.org/pdf_library/data_tables/econ2_2003.pdf (accessed September 1, 2003).

———. 2003b. "Financial Flows, Government Expenditures, and Corporations 2003." www.earthtrends.wri.org/pdf_library/data_tables/gov3_2003.pdf (accessed September 1, 2003).

Yale Center for Environmental Law and Policy. 2002. *Environmental Sustainability Index*. www.ciesin.columbia.edu/indicators/ESI.

Young, Emma. 2004. "Taboos could Save the Seas." *New Scientist* 182 (2443): 9.

Zimmerer, Karl, and Eric Carter. 2001. "Conservation and Sustainability in Latin America and Caribbean." In *Latin America in the 21st Century*. Austin: University of Texas Press.

INDEX

AES Corp., 121, 134, 142
Afghanistan, 97
Africa, 20, 69, 89, 112
AIDCP. *See* International Dolphin Conservation Program
Al-Qaeda, 97
American Samoa, 65, 67, 76–77
Antigua and Barbuda, 87, 99, 107
Apa, Martin, killing, 79
Arawak people. *See* Taino people
Aristide, Jean-Bertrand, 88, 97
Asia, 20, 73, 85, 99, 102, 119–21, 126–27; ASEAN, 81, 106, 114–15, 120, 122, 130–32, 139, 146; ASEAN Regional Center for Biodiversity, 129, 131; Fisheries Development Center Aquaculture Department 129; Southeast Asia, 15, 24, 31, 66–67, 74–75, 79, 87, 92, 99, 109, 111–35, 138, 142, 146, 151–52, 155, 157, 162–63, 165. *See also* tsunami
Asian Coalition for Housing, 132
Asian Development Bank (ADB), 73–74, 79, 113, 116, 121, 124–26, 128–29, 147
Atlantic bluefin tuna. *See* tuna
Atlantic Ocean, 9, 39–40, 42, 52, 70, 95, 158
Australia, 11, 50, 55, 56, 65–69, 74–76, 80, 83, 86, 148, 151; AusAid, 79

backdown, dolphin preservation tactic of, 9
Bahamas, 87, 90, 99
Bangladesh, 12, 58, 112–15, 120, 125, 164

Barbados, 87, 99, 104, 106
Belize, 87, 92, 94, 98
Bello, Walden, 132
Black Sea, 41
bleaching, coral, 50, 52, 96
"boat people," 132, 165
Borgese, Elisabeth Mann, 10–12, 14, 19–20, 29, 33–34, 62–63, 75, 86n1, 145, 150, 159, 161–62, 165
Bougainville, 75
Britain. *See* United Kingdom
Brundtland Report, 14
Brunei, 56, 112, 114–15, 131
Bumble Bee Tuna, 76
Burma, 111–15, 122, 131

Cambodia, 58, 112, 114–15, 120, 124, 131
Cargill Corporation, 125, 142
Carib people, 89, 105
Caribbean, 15, 24, 31, 34, 49, 59, 66–68, 74, 79, 87–110, 113, 117, 121, 133, 138–39, 142, 151, 152, 156, 162, 164; CARICOM, 81, 87, 93, 99–100, 106–09, 139, 146; Common Market, 106; Conservation Corporation, 104; Fisheries Resource Assessment and Management Program, 99; Latin American Action, 106; Latin American Caribbean region (LAC), 67–68, 88, 98, 102; Natural Resource Institute (CANARI), 104, 109, 142; Network of Women Producers, 105; St. George's Declaration of Princi-

185

ples for Environmental Sustainability, 107; West Indies, University of, 108
Carnival Cruise Lines, 91
Carson, Rachel, 18
Casanova, Jose Manuel, 104
Caspian Sea, 40
catch per unit of effort, 23–24, 75
Challenger, H.M.S., 62
Charoen Pokphand Group (CP Group) (Thailand) 15, 123–24, 126
Chile, 12
China, 15, 46, 54–55, 58, 111–12, 114–16, 120–21, 123, 135, 151, 164
Citibank, 142
Colombia, 88, 92, 94, 97–98, 121
comanagement, 11, 80, 99, 109, 134, 141, 155
Confucianism, 132–33
conservation biology, 9
Continental Grain Corporation, 123
Cook Islands, 57, 65, 67
Costa Rica, 88, 92, 94, 98
crown of thorns starfish, 50, 71
Cuba, 87–8, 90, 98, 100, 104, 108

Dark Ages, 162
dasein, 27–29, 145
Denmark, 89
distancing or distanciation, 6
Dominican Republic, 57, 87–88, 90–92, 94–95, 98, 105, 107
Down to Earth environmental group, 126, 142
Dutch East India Company, 112

East Asian Sea, 127
East Timor, 112, 129, 132
Easter Island, 65
ecosemiotics, 148
Egypt, 58
"end of history," 2
entropy and the economic problem, 7
El Niño Southern Oscillation (ENSO), 70, 73, 118, 149

El Salvador, 88, 103
European Union (EU), 3–5, 78, 90, 107, 121, 130–31, 135
exclusive economic zones (EEZs), 45, 73, 77–78, 154

Federated States of Micronesia (FSM), 65, 67, 76
Fiji, 65, 68, 72, 74–75
Fishery Stock Agreement (FSA), 63, 82, 161
flags of convenience, 20, 91
Fonseca, Committee for the Defense and Development of the Flora and Fauna of the Gulf of, 102, 110
Food and Agriculture Organization (FAO), 41, 46, 66, 69, 70–71, 87, 95, 114, 116–17, 133, 138, 164
foreign direct investment (FDI), 35, 67–69, 83, 85, 92–93, 113–14, 116, 120, 135, 151–54
Forum Fisheries Agency. *See* South Pacific
France, 55, 66–67, 76, 82–83, 89
Frankfurt School, 29, 150
French Polynesia, 65, 67
Fuerzas Armadas Revolucionaris Colombians (Revolutionary Armed Forces of Colombia), 97–98

Gandhi, Mahatma, 10, 150, 159
Global Coral Reef Monitoring Network, 12
Global Ocean Observing System, 12
globalization, 1–9, 11–12, 15–16, 19–23, 29–35, 37, 39, 43, 46, 61, 63, 66–68, 75–76, 81–86, 88–91, 101–2, 106, 108–10, 112, 114, 120, 123, 135, 142–43, 145–46, 150, 153, 155–57, 163–64; anti-, 132; globalism, 22
Great Barrier Reef, 78
Greenland Ice Sheet, 55
Grenada, 57, 87, 92, 107
Grotius, Hugo, 60
Guam, 56, 65–67
Guatemala, 88, 92, 94, 98, 103
Guyana, 87, 92, 94, 98, 106

Hae, Lee Kyoung, 5
Haiti, 87–88, 90, 92, 94, 97–98, 108, 146
Havana, University of, 100
Hegel, Friedrich, 142, historical materialism, 142, 158
Honduras, 88, 92, 94, 98, 102–3, 125
Hussein, Saddam, 97

Indian Ocean, 39, 58, 69, 114, 116–19
Indonesia, 55, 75, 111–12, 114–15, 117–18, 120–26, 129, 131; Aceh, 113, 124; Bali, 113
Industrial Revolution, 51
Inter-American Tropical Tuna Commission (IATTC), 9
International Basic Economy Corporations, 123
International Coral Reef Action Network, 117–18
International Dolphin Conservation Program (AIDCP), 9
International Monetary Fund (IMF), 5–6, 106, 110, 113, 120–21, 150
International Panel on Climate Change, 53
International Seabed Authority, 63
Iraq, 97
IUCN. *See* World Conservation Union

Jamaica, 49, 63, 87, 92, 94–95, 98, 106, 152
Japan, 20, 55, 70, 76–78, 102, 111, 115, 121, 126, 135, 141, 148, 151, 155, 157; government, 121. *See also* Tsukiji Fish Market
Japfa Comfeed Corporation, 125
Johannes, Robert, 79–80

Kentucky Fried Chicken Corporation, 123
Kiribati, 35, 57, 65, 78, 117
Korea, Democratic People's Republic of (North Korea), 55
Korea, Republic of (South Korea), 55, 76–77, 121, 132

La Niña event, 70, 118
Laos, 111, 113–15, 120, 128–29
Law of the Sea (UNCLOS), 2, 10, 19, 33, 60–63, 154; Common Heritage for Mankind, 2, 10, 33, 59, 62–63
liberalism, 2, 4–5, 74, 84–85, 101, 103, 120, 146–48, 150–51; neoliberalism, 4–8, 11, 16, 21, 47–48, 61, 74, 81, 83–84, 90, 101–3, 105, 109–10, 113–14, 126, 128, 130, 132–33, 135, 141, 146–47, 149, 154, 164
Lome Convention, 90

Malaysia, 15, 79, 111–12, 114–15, 120–22, 124–25, 155
Maldives, 58
Mangrove Action Project, 150
mare liberum, 43, 59–61, 63, 161. *See also* Grotius, Hugo
marine protected areas (MPAs), 15, 100, 104, 131, 134
Marshall Islands, 65–67
Marx, Karl, 142
material flow accounting, 129
Mexico, 5, 55, 87–88, 92, 94, 98, 103
Mitsubishi, 76, 123
modernity, 4, 30–31, 103
Monroe Doctrine, 80, 90
Montserrat, 57, 87, 107

Nature Conservancy, 101
Nauru, 56–57, 66, 86
neoliberalism. *See* liberalism
New Caledonia, 66–67, 75
New York City, 20
New Zealand, 11, 66–68, 71, 74, 76, 80
Newtonian mechanism, 25
Nigeria, 58
Niue, 57, 66–67
North America, 20, 50, 67, 91, 148
North American Treaty Organization (NATO), 3
Northern Mariana Islands (CNMI), 66–67
nucleus estate aquaculture regime, 126

Oceana, environmental group, 91
orange roughy, 70
Organization of Eastern Caribbean States (OECS), 99, 107–8
Organization for Economic Cooperation and Development (OECD), 2, 4
Oxfam, 15, 122

Pacific Ocean, 12, 39–40, 66–67, 70–73, 77–78, 82, 117, 119, 122, 165
Palau, 56, 66–67, 86
Panama, 88, 92, 94, 98, 103
Panday, Oma, 105
Papua New Guinea (PNG), 66, 68, 74–76, 78, 79, 84
Pardoe, Arvid, 10, 62–63
Philippines, 50, 76, 102, 111–15, 117, 120–22, 125, 129, 134
Phosphate Chemicals Export Association, 125
Pitcairn Islands, 66–67
Polynesia. *See* French Polynesia
Potash Corporation, 125
Princess Cruise Lines, 91
problematiques, 11
Puerto Rico, 76, 87

Rastafarians, 105
Reagan, Ronald President, 2
Regional Seas Program, United Nations Environmental Programme, 34
Roratonga, Treaty of, 82
Royal Caribbean International, 91

Samoa, 57, 66, 74, 79. *See also* American Samoa
"sea gypsies". *See* "boat people"
SEAfish for Justice, 122
Seward, William, Secretary of State, 90
Singapore, 56, 112, 114–15, 120, 127, 131
Solomon Islands, 57, 66, 68, 74–76, 78–79
solutiques, 11
South Korean farmer who committed suicide at WTO protest. *See* Hae, Lee Kyoung

South Pacific, 11, 15, 20, 31, 34, 36–37, 58, 65–86, 87, 92, 99, 101, 104, 106, 111, 113, 117, 121–22, 133, 135, 138–39, 141, 145–46, 151–52, 155, 161–62, 165; Forum Fisheries Agency (FFA), 76, 78, 81–82; Pacific Women's Bureau, 84; Regional Environmental Programme (SPREP), 81–3, 145–46; South Pacific Nuclear Free Zone Treaty (*see* Roratonga, treaty of);
South Pacific Secretariat, 66, 70, 81, 145–46; University of, 81
Southeast Asia. *See* Asia
Southern African Development Community (SADC), 130
Southern Ocean, 39, 52, 140
Soviet Union, 3, 78, 88, 113
Spain, 55, 59, 66, 76, 89
Sri Lanka, 112, 114–15, 120, 125
St. Kitts and Nevis, 57, 87
St. Lucia, 87, 102, 109
St. Vincent and the Grenadines, 57, 87, 98
subaltern, 31
Suriname, 87, 94

Taino people, 89
Taiwan, 55, 76–77, 124, 132
Tamil Nadu Gram Swaraj Movement, 150
Thai Union Frozen Products, 76
Thailand, 15, 55, 76–77, 112, 114–15, 120–25, 127–29, 131; Bank of, 124; Gulf of, 123
things in themselves. *See dasein*
Third World Network, 15
throughput, 5–7
Tokelau, 66–67, 153
Tong, Sony, killing, 79
Tonga, 57, 66, 74, 79
treadmills of production, 30
Trinidad and Tobago, 56, 87, 92, 94, 97–98, 101, 105–6
Tsukiji Fish Market, 20
tsunami, of December 24, 2004, 124

tuna, 9, 20, 35, 41, 63, 67, 70, 73, 75–78, 81, 85–86, 95, 123, 153, 164
Tuvalu, 58, 66, 77

United Kingdom, 4, 55, 66–67, 82, 89, 106, 112–13, 126
United Nations Conference on Highly Migratory and Straddling Fish Stocks. *See* Fishery Stock Agreement
United Nations Environmental Programme, 34, 81
United Nations Food and Agriculture Organization. *See* Food and Agriculture Organization
United Nations Framework on Climate Change, 82
United Nations Law of the Sea Conferences. *See* Law of the Sea
United States, 2, 5, 12, 29–30, 35, 54–56, 59, 66–69, 76, 78, 82–83, 87–92, 94, 97–98, 101–2, 104, 106, 108, 111, 113, 115, 117, 121, 122–23, 125–26, 131, 132, 135, 139, 141, 148, 150–51, 155–57, 164; Agency for International Development (USAID), 102, 142

Van Camp Seafood, 76
Vanatinai, 84
Venezuela, 88, 92, 94, 98, 109
Vietnam, 112–15, 117, 120–22, 124, 126, 129, 131

Wallis and Futuna, 66–67
Washington Consensus, 146, 162
West Indian Federation, 106
Western Central Atlantic Fisheries Commission, 87
World Bank, 2–5, 16, 36, 67, 73, 75, 77, 79, 113–15, 121, 124–26, 150–51
world capitalist system, 47, 141, 150, 158, 162
World Conservation Union (IUCN), 79, 101; IUCN Red List, 134
World Fish Center, 117
World Ocean Circulation Experiment, 12
World Rainforest Movement environmental group, 126, 142
World Trade Organization (WTO), 3, 5
World War I, 2
World War II, 2, 46, 66, 112
World Wide Fund for Nature (WWF), 15

About the Author

Peter Jacques is assistant professor in the Department of Political Science at the University of Central Florida in Orlando, where he teaches global environmental politics, sustainability, urban environmental politics, and domestic environmental politics to graduate and undergraduate students. He received his PhD at Northern Arizona University in 2003 in domestic and global environmental politics. He is coauthor of *Ocean Politics and Policy: A Reference Handbook* (2003) with Zachary Smith, and has published articles on environmental politics in several journals, including *Global Environmental Politics*, *The Social Science Journal*, *The Radical Review of Political Economy*, and others.

www.ingramcontent.com/pod-product-compliance
Ingram Content Group UK Ltd.
Pitfield, Milton Keynes, MK11 3LW, UK
UKHW021902220326
469204UK00008B/138